Selected Titles in This Series

710 **Brian Marcus and Selim Tuncel,** Resolving Markov chains onto Bernoulli shifts via positive polynomials, 2001

709 **B. V. Rajarama Bhat,** Cocylces of CCR flows, 2001

708 **William M. Kantor and Ákos Seress,** Black box classical groups, 2001

707 **Henning Krause,** The spectrum of a module category, 2001

706 **Jonathan Brundan, Richard Dipper, and Alexander Kleshchev,** Quantum Linear groups and representations of $GL_n(\mathbb{F}_q)$, 2001

705 **I. Moerdijk and J. J. C. Vermeulen,** Proper maps of toposes, 2000

704 **Jeff Hooper, Victor Snaith, and Min van Tran,** The second Chinburg conjecture for quaternion fields, 2000

703 **Erik Guentner, Nigel Higson, and Jody Trout,** Equivariant E-theory for C^*-algebras, 2000

702 **Ilijas Farah,** Analytic guotients: Theory of liftings for quotients over analytic ideals on the integers, 2000

701 **Paul Selick and Jie Wu,** On natural coalgebra decompositions of tensor algebras and loop suspensions, 2000

700 **Vicente Cortés,** A new construction of homogeneous quaternionic manifolds and related geometric structures, 2000

699 **Alexander Fel′shtyn,** Dynamical zeta functions, Nielsen theory and Reidemeister torsion, 2000

698 **Andrew R. Kustin,** Complexes associated to two vectors and a rectangular matrix, 2000

697 **Deguang Han and David R. Larson,** Frames, bases and group representations, 2000

696 **Donald J. Estep, Mats G. Larson, and Roy D. Williams,** Estimating the error of numerical solutions of systems of reaction-diffusion equations, 2000

695 **Vitaly Bergelson and Randall McCutcheon,** An ergodic IP polynomial Szemerédi theorem, 2000

694 **Alberto Bressan, Graziano Crasta, and Benedetto Piccoli,** Well-posedness of the Cauchy problem for $n \times n$ systems of conservation laws, 2000

693 **Doug Pickrell,** Invariant measures for unitary groups associated to Kac-Moody Lie algebras, 2000

692 **Mara D. Neusel,** Inverse invariant theory and Steenrod operations, 2000

691 **Bruce Hughes and Stratos Prassidis,** Control and relaxation over the circle, 2000

690 **Robert Rumely, Chi Fong Lau, and Robert Varley,** Existence of the sectional capacity, 2000

689 **M. A. Dickmann and F. Miraglia,** Special groups: Boolean-theoretic methods in the theory of quadratic forms, 2000

688 **Piotr Hajłasz and Pekka Koskela,** Sobolev met Poincaré, 2000

687 **Guy David and Stephen Semmes,** Uniform rectifiability and quasiminimizing sets of arbitrary codimension, 2000

686 **L. Gaunce Lewis, Jr.,** Splitting theorems for certain equivariant spectra, 2000

685 **Jean-Luc Joly, Guy Metivier, and Jeffrey Rauch,** Caustics for dissipative semilinear oscillations, 2000

684 **Harvey I. Blau, Bangteng Xu, Z. Arad, E. Fisman, V. Miloslavsky, and M. Muzychuk,** Homogeneous integral table algebras of degree three: A trilogy, 2000

683 **Serge Bouc,** Non-additive exact functors and tensor induction for Mackey functors, 2000

682 **Martin Majewski,** ational homotopical models and uniqueness, 2000

681 **David P. Blecher, Paul S. Muhly, and Vern I. Paulsen,** Categories of operator modules (Morita equivalence and projective modules, 2000

(Continued in the back of this publication)

Resolving Markov Chains onto Bernoulli Shifts via Positive Polynomials

Memoirs
of the
American Mathematical Society

Number 710

Resolving Markov Chains
onto Bernoulli Shifts via
Positive Polynomials

Brian Marcus
Selim Tuncel

March 2001 • Volume 150 • Number 710 (first of 5 numbers) • ISSN 0065-9266

American Mathematical Society
Providence, Rhode Island

2000 *Mathematics Subject Classification.*
Primary 28D20, 11C08; Secondary 05A10, 94A17.

Library of Congress Cataloging-in-Publication Data
Marcus, Brian, 1949–
  Resolving Markov chains onto Bernoulli shifts via positive polynomials / Brian Marcus, Selim Tuncel.
     p. cm. — (Memoirs of the American Mathematical Society, ISSN 0065-9266 ; no. 710.)
  "Volume 150, number 710 (first of 5 numbers).
  Includes bibliographical references.
  ISBN 0-8218-2646-8
  1. Markov processes. 2. Bernoulli shifts. I. Tuncel, Selim. II. Title. III. Series.
QA3.A57 no. 710
[QA274.7]
510s—dc21
[519.2'33]   00-053581

Memoirs of the American Mathematical Society

This journal is devoted entirely to research in pure and applied mathematics.

Subscription information. The 2001 subscription begins with volume 149 and consists of six mailings, each containing one or more numbers. Subscription prices for 2001 are $494 list, $395 institutional member. A late charge of 10% of the subscription price will be imposed on orders received from nonmembers after January 1 of the subscription year. Subscribers outside the United States and India must pay a postage surcharge of $31; subscribers in India must pay a postage surcharge of $43. Expedited delivery to destinations in North America $35; elsewhere $130. Each number may be ordered separately; *please specify number* when ordering an individual number. For prices and titles of recently released numbers, see the New Publications sections of the *Notices of the American Mathematical Society*.

Back number information. For back issues see the *AMS Catalog of Publications*.
Subscriptions and orders should be addressed to the American Mathematical Society, P. O. Box 845904, Boston, MA 02284-5904. *All orders must be accompanied by payment.* Other correspondence should be addressed to Box 6248, Providence, RI 02940-6248.

Copying and reprinting. Individual readers of this publication, and nonprofit libraries acting for them, are permitted to make fair use of the material, such as to copy a chapter for use in teaching or research. Permission is granted to quote brief passages from this publication in reviews, provided the customary acknowledgment of the source is given.

Republication, systematic copying, or multiple reproduction of any material in this publication is permitted only under license from the American Mathematical Society. Requests for such permission should be addressed to the Assistant to the Publisher, American Mathematical Society, P. O. Box 6248, Providence, Rhode Island 02940-6248. Requests can also be made by e-mail to reprint-permission@ams.org.

Memoirs of the American Mathematical Society is published bimonthly (each volume consisting usually of more than one number) by the American Mathematical Society at 201 Charles Street, Providence, RI 02904-2294. Periodicals postage paid at Providence, RI. Postmaster: Send address changes to Memoirs, American Mathematical Society, P. O. Box 6248, Providence, RI 02940-6248.

© 2001 by the American Mathematical Society. All rights reserved.
This publication is indexed in *Science Citation Index*®, *SciSearch*®, *Research Alert*®, *CompuMath Citation Index*®, *Current Contents*®/*Physical, Chemical & Earth Sciences*.
Printed in the United States of America.

∞ The paper used in this book is acid-free and falls within the guidelines
established to ensure permanence and durability.
Visit the AMS home page at URL: http://www.ams.org/

10 9 8 7 6 5 4 3 2 1   06 05 04 03 02 01

Contents

Preface ix

Part A.
Resolving Markov Chains onto Bernoulli Shifts 1

1. Introduction 3
2. Weighted graphs and polynomial matrices 6
3. The main results 11
4. Markov chains and regular isomorphism 18
5. Necessity of the conditions 25
6. Totally conforming eigenvectors and the one-variable case 31
7. Splitting the conforming eigenvector in the one-variable case 40
8. Totally conforming eigenvectors for the general case 53
9. Splitting the conforming eigenvector in the general case 58

Bibliography 72

Part B.
On Large Powers of Positive Polynomials in Several Variables 75

1. Introduction 77
2. Structure of $\mathbf{Log}(\mathbf{p}^n)$ 80
3. Entropy and equilibrium distributions for $\mathbf{w} \in \mathbf{W}(\mathbf{p})$ 85
4. Equilibrium distributions and coefficients of $\mathbf{p}^n$ 90
5. Proofs of the estimates 93

Bibliography 98

Preface

The two parts of this Memoir contain two separate but related papers. The longer paper in Part A obtains necessary and sufficient conditions for several types of codings of Markov chains onto Bernoulli shifts. It proceeds by replacing the defining stochastic matrix of each Markov chain by a matrix whose entries are polynomials with positive coefficients in several variables; a Bernoulli shift is represented by a single polynomial with positive coefficients, p. This transforms jointly topological and measure-theoretic coding problems into combinatorial ones. In solving the combinatorial problems in Part A, we state and make use of facts from Part B concerning p^n and its coefficients.

Part B contains the shorter paper on p^n and its coefficients, and is independent of Part A.

An announcement describing the contents of this Memoir may be found in the Electronic Research Announcements of the AMS at the following Web address: http://www.ams.org/era/

Received by the editor February 8, 1999.

The second author was partially supported by NSF Grant DMS-9622866 and also thanks Erwin Schroedinger Institute, Vienna for their support and hospitality during a visit in the summer of 1997.

Part A

Resolving Markov Chains onto Bernoulli Shifts

Introduction

It is well known from Ornstein's isomorphism theory [**FO, O**] that entropy is a complete invariant of measure-theoretic isomorphism for mixing Markov chains. In that result the isomorphism is a *two-sided isomorphism*, meaning that the isomorphism and its inverse are allowed to use the entire future and the entire past of a point in order to determine the zeroth coordinate of its image. In a *one-sided isomorphism*, the isomorphism and its inverse are allowed to use the entire past, but none of the future. In [**AMT**] Markov chains were completely classified up to one-sided measure-theoretic isomorphism, via an effective algorithmic procedure.

An intermediate notion of isomorphism is *regular isomorphism*. Here, the isomorphism and its inverse are allowed to use the entire past and a uniformly bounded amount of the future. Regular isomorphisms were introduced in [**FP**] and further studied by Parry and others (see [**PT3, BT**] and their references). One of our results will give necessary and sufficient conditions for a Markov chain to be regularly isomorphic to a Bernoulli shift.

Ideas emanating from symbolic dynamics, in particular right-resolving and right-closing maps of shifts of finite type (see, for example, [**LM**]), have a strong bearing on one-sided isomorphism and regular isomorphism of Markov chains: If a right-closing factor map is one-to-one almost everywhere then it defines a (rather concrete) regular isomorphism between the shifts of finite type with respect to their measures of maximal entropy. More generally, if the shifts of finite type are endowed by Markov measures and the map preserves measure, then we obtain a regular isomorphism from one of the Markov chains to the other. It was shown in [**BT**] that two Markov chains are regularly isomorphic if and only if there exists a Markov chain which factors onto each of them by a right-closing factor map which is one-to-one almost everywhere. Thus, the study of regular isomorphism for Markov chains largely reduces to the study of right-closing factor maps between Markov chains.

Likewise, two Markov chains are one-sidedly isomorphic if and only if there exists a Markov chain which factors onto each of them by a right-resolving map which is one-to-one almost everywhere. This fact serves as the first step of the classification of [**AMT**]. Right-closing maps form the closure of right-resolving ones under block isomorphism, that is, measure-preserving topological conjugacy.

In this paper, we will completely answer the following questions for an arbitrary Markov chain and a Bernoulli shift.

(1) Does the Markov chain eventually factor onto the Bernoulli shift by right-closing maps?
  (2) Does the Markov chain factor onto the Bernoulli shift by a right-closing map? By a right-closing map of degree 1?
  (3) Is the Markov chain regularly isomorphic to the Bernoulli shift?

Without consideration of measures or, equivalently, restricting attention to measures of maximal entropy, the analogues of questions (1)–(3) were answered for shifts of finite type in the sequence of papers [**M, T, BMT, A2**] in terms of topological entropy, dimension groups and the factor periodic point condition. It is natural to seek generalizations of these results to Markov chains. In this paper the generalizations are carried out for a Bernoulli image.

Our solutions are facilitated by nonnegative matrices of polynomials in several variables which are associated to Markov chains along the lines of [**PT2, T2, MT1, T3**]. We have, thus, a nonnegative polynomial matrix A and a nonnegative polynomial p, representing a Markov chain and a Bernoulli shift. The matrix A defines a weighted graph $G(A)$. We denote by (X_A, σ_A) the resulting shift of finite type together with the weights-per-symbol assigned to its periodic orbits. The necessary and sufficient conditions for (1)–(3) involve only the basic invariants $\beta, \Delta, c\Delta$, the factor periodic point condition and a right eigenvector r of A. In this context the factor periodic condition takes the following form: For each periodic orbit γ of (X_A, σ_A) there exists in (X_p, σ_p) a periodic orbit of the same weight-per-symbol whose period divides that of γ.

The right eigenvector is required to satisfy three conditions, which we label (a)–(c). Condition (a) asks that the entries of r be polynomials with positive coefficients. The existence of an eigenvector with this property was demonstrated in [**MT3**]. Conditions (b) and (c) are extracted with the aid of the faces $\mathcal{F}(p)$ of the Newton polyhedron $W(p)$ of p in a manner closely related to [**MT1, T3**]. Let $F \in \mathcal{F}(p)$. We have, as in [**H1, MT3, T3**], the F-face p_F of p and the F-components of A. Condition (b) may be stated roughly as follows. Suppose F is a proper face of $W(p)$, and let I be a state of an F-component of A. The corresponding entry $r(I)$ must have a well-defined F-face $r(I)_F$ and, for large n, we must be able to translate the support $\text{Log}(r(I))$ of $r(I)$ by a vector u so that $\text{Log}(r(I)_F) + u \subset \text{Log}(p_F^n)$ and, simultaneously, $\text{Log}(r(I)) + u \subset \text{Log}(p)$; we will characterize this situation by saying that $r(I)$ must F-conform to p. Condition (c) requires that for each $F \in \mathcal{F}(p)$ and every period class C of each principal F-component of A, the set $\{r(I)_F : I \in C\}$ generate (over the appropriate coefficient ring) the entire dimension module of p_F. In view of [**T2, MT1**], (c) is the proper generalization of the dimension condition of [**BMT, A2**] to our context. It consists of a global module condition and a finite list of analogous boundary module conditions, one for each principal facial component of A. In terms of polynomial matrices and weight-preserving maps, our answers to (1)–(3) may be stated as follows.

  (1) (X_p, σ_p) is a right-closing eventual factor of (X_A, σ_A) if and only if $\beta_A = p$ and there exists a right eigenvector r of A satisfying (a)–(c).

(2) (X_p, σ_p) is a right-closing factor of (X_A, σ_A) if and only if $\beta_A = p$, there exists a right eigenvector r of A satisfying (a)–(c), and the factor periodic point condition holds. Moreover, (X_p, σ_p) is a factor of (X_A, σ_A) by a right-closing map of degree 1 if and only if, in addition, A is aperiodic, $\Delta(A) = \Delta(p)$ and $c_A \Delta(A) = c_p \Delta(p)$.

(3) The Markov chain represented by (X_A, σ_A) is regularly isomorphic to the Bernoulli shift represented by (X_p, σ_p) if and only if A is aperiodic, $\beta_A = p$, $\Delta(A) = \Delta(p)$, $c_A \Delta(A) = c_p \Delta(p)$ and there exists a right eigenvector r of A satisfying (a)–(c).

The terminology used in the above description and the conditions (a)–(c) will be explained in detail in sections 2 and 3. For the time being, we remark that (b) goes beyond formulating the boundary counterpart of a global condition; it is a near-boundary condition. Near-boundary obstructions have been observed (in [MT2, T3] and in unpublished work of the authors) to arise in a variety of coding problems for Markov chains. The results of this paper are the first instances of the proper and full formulation of such constraints in solving coding problems.

Our main results will be stated in sections 3 and 4 as theorems (3.2), (3.3), (4.1), (4.2) and (4.3), and their proofs will take up sections 4–9. The results of [MT4] on the large powers p^n of p will be crucial ingredients. One of these results, stated below as (3.6), describes the support $\text{Log}(p^n)$ for all large n. The others, stated below as (7.1)–(7.3) and (9.1)–(9.3), concern relative sizes of the coefficients of p^n for large n.

The paper is organized as follows. With the exception of section 4, we work in the setting of matrices of nonnegative polynomials in several variables. In section 2 we introduce the basic objects and tools. Section 3 contains the statements of our main results in the context of polynomial matrices and a detailed discussion of their conditions. In section 4 we explain how to associate nonnegative matrices of polynomials to Markov chains, and thus answer (1)–(3). In fact, using [A1, A2, BT, MT4], the solution of (3) may be obtained from that of (2), which in turn follows from that of (1); these observations are explained in sections 4 and 5, respectively. With section 5 we return to polynomial matrices and their weighted graphs: We establish the relationship between right-closing maps and certain state-splittings (partitions of paths) in the weighted graph $G(A)$ associated to the matrix A. We also demonstrate the necessity of our conditions for (1). Sections 6–9 of the paper are devoted to the sufficiency of these conditions for right-closing eventual factoring. This requires some rather involved state-splittings. We first consider, in sections 6 and 7, the case of polynomials of a single variable, which allows us to present the basic ideas of the proof without the additional complication of inductions based on faces of $W(p)$. We carry out two state-splittings. The first one produces a totally conforming eigenvector (every entry is F-conforming for every face F of $W(p)$), which helps with the second splitting. Using the information provided by [MT4], sections 8 and 9 then build the inductions for implementing the state-splittings in the general case.

Weighted graphs and polynomial matrices

Let G be a directed graph, denote its set of states (vertices) by $S(G)$ and its set of edges by $E(G)$. For $e \in E(G)$ we let $s(e)$ denote the starting state of e, and $t(e)$ its terminal state. Putting

$$X_G = \{(e_n)_{n \in \mathbb{Z}} : e_n \in E(G), \, t(e_n) = s(e_{n+1}) \, \forall n \in \mathbb{Z}\},$$

we have the shift of finite type (X_G, σ_G). For directed graphs, shifts of finite type and their block maps, we take [**LM**] as the standard reference and make free use of the definitions, notation and basic facts therein. In particular, see [**LM**] for definitions and basic facts related to factor maps (codes), finite-to-one codes and their degrees, right-resolving maps, and right-closing maps. Whenever we deviate from the terminology or notation of [**LM**], we give explicit definitions. For example, according to the definitions given above, we denote the starting state of e by $s(e)$ and thus deviate from the notation $i(e)$ used in [**LM**]. Letting H be a second directed graph, recall that if $\phi : X_G \to X_H$ is a continuous map with $\phi \sigma_G = \sigma_H \phi$ then there exist $k \in \mathbb{Z}^+$, $l \in \mathbb{Z}$ and a map $\Phi : B_k(X_G) \to S(H)$ of the k-blocks of X_G such that

$$\phi(x)_n = \Phi(x_{n+l} \, x_{n+l+1} \cdots x_{n+l+k-1})$$

for all $n \in \mathbb{Z}$; we then say that ϕ is a k-*block map* and that it has *lag l*, and also write ϕ for Φ. A pair of maps $\phi : E(G) \to E(H)$, $\phi_0 : S(G) \to S(H)$ defines a *graph homomorphism* if $s(\phi(e)) = \phi_0(s(e))$ and $t(\phi(e)) = \phi_0(t(e))$ for all $e \in E(G)$. We denote this by $(\phi, \phi_0) : G \to H$. A graph homomorphism $(\phi, \phi_0) : G \to H$ is *right-resolving* if ϕ_0 is surjective and, for each $I \in S(G)$, the restriction of ϕ to $\{e \in E(G) : s(e) = I\}$ is a bijection of this set onto $\{f \in E(H) : s(f) = \phi_0(I)\}$. Clearly, 1-block maps $\phi : X_G \to X_H$ are obtained from graph homomorphisms $(\phi, \phi_0) : G \to H$, and right-resolving factor maps are obtained from right-resolving graph homomorphisms.

The directed graph G is *weighted* if, for some multiplicative Abelian group $\mathcal{G}$, each $e \in E(G)$ is assigned a *weight* $\operatorname{wt}(e) = \operatorname{wt}_G(e) \in \mathcal{G}$. In this case, the *weight* of a G-path $\gamma = e_1 e_2 \cdots e_n$ is the product

$$\operatorname{wt}(\gamma) = \operatorname{wt}_G(\gamma) = \operatorname{wt}(e_1) \operatorname{wt}(e_2) \cdots \operatorname{wt}(e_n).$$

Following [**K2, T2, PS**], we define $\Gamma(G)$ to be the subgroup of $\mathcal{G}$ generated by weights of cycles and let

$$\Delta(G) = \{\operatorname{wt}(\gamma)/\operatorname{wt}(\eta) : \gamma \text{ and } \eta \text{ are } G\text{-cycles of the same length}\}.$$

It is easy to see that $\Gamma(G)$ is finitely generated and $\Delta(G)$ is a subgroup of $\Gamma(G)$. In the case G is irreducible and has period d, we let γ and η be G-cycles whose lengths satisfy $l(\gamma) = l(\eta) + d$ and use $c_G = \operatorname{wt}(\gamma)/\operatorname{wt}(\eta)$ to distinguish a coset $c_G \Delta(G)$ of $\Delta(G)$. It is easy to check that $c_G \Delta(G)$ is independent of the choice of cycles γ, η with $l(\gamma) = l(\eta) + d$. (For details and further properties of $\Gamma, \Delta, c\Delta$, see [**PS, MT1**].) The graphs we work with will always be directed and, with the exception of section 4, weighted by monomials.

Let $R_k = \mathbb{Z}[x_1^\pm, x_2^\pm, \ldots, x_k^\pm]$ be the ring of Laurent polynomials in k variables and $R_k^+ = \mathbb{Z}^+[x_1^\pm, \ldots, x_k^\pm]$ its positive cone. For $w = (w_1, \ldots, w_k) \in \mathbb{Z}^k$, we put

$$x^w = x_1^{w_1} x_2^{w_2} \cdots x_k^{w_k}.$$

The map $w \mapsto x^w$ gives a group isomorphism between $\mathbb{Z}^k$ and the multiplicative group of monomials in R_k, allowing us to move between these two groups. Let $p \in R_k$. We denote the coefficient of x^w in p by p_w. Then $p_w \in \mathbb{Z}$,

$$p = \sum_{w \in \mathbb{Z}^k} p_w x^w,$$

and p_w are nonzero for only finitely many $w \in \mathbb{Z}^k$. We define

$$\operatorname{Log}(p) = \{w \in \mathbb{Z}^k : p_w \neq 0\}.$$

The *Newton polyhedron* of p, denoted by $W(p)$, is the convex hull of $\operatorname{Log}(p)$ in $\mathbb{Q}^k$. We denote the collection of nontrivial faces of $W(p)$ by $\mathcal{F}(p)$. For $F \in \mathcal{F}(p)$, we let

$$p_F = \sum_{w \in F} p_w x^w,$$

and call p_F the *F-face* of p. For a set Δ of monomials in R_k, we put

$$\operatorname{Log}(\Delta) = \{w \in \mathbb{Z}^k : x^w \in \Delta\}.$$

For a group Δ of monomials in R_k, we let $R(\Delta)$ denote the integral group ring of Δ; that is,

$$R(\Delta) = \{q \in R_k : \operatorname{Log}(q) \subset \operatorname{Log}(\Delta)\}.$$

The positive cone of $R(\Delta)$ is $R(\Delta)^+ = R_k^+ \cap R(\Delta)$.

Let A be an irreducible matrix over R_k^+. We associate to A a weighted graph $G(A)$ with the indexing set of A as its states: We express each entry $A(I, J)$ as a sum (with multiplicity) of monomials, place as many edges from I to J as there are terms in this sum, and bijectively assign the monomials to these edges as weights. This gives us a one-to-one correspondence between matrices over R_k^+ and graphs weighted by monomials since, given such a graph G, we can index a matrix A by $S(G)$ and, for $I, J \in S(G)$, let $A(I, J)$ equal the sum of the weights of the edges from I to J. We shall allow ourselves to confuse A with $G(A)$ and write $\operatorname{wt}_A(e), \operatorname{wt}_A(\gamma), \Gamma(A), \Delta(A), c_A \Delta(A)$ for the objects defined above for $G(A)$. For $G(A)$-cycle γ of length $l = l(\gamma)$, we define the *weight-per-symbol* of γ to be

$$\operatorname{wps}_A(\gamma) = \frac{1}{l} \operatorname{Log}(\operatorname{wt}_A(\gamma)) \in \mathbb{Q}^k.$$

We write (X_A, σ_A) for the shift of finite type defined by the unweighted graph underlying $G(A)$ together with the weights-per-symbol assigned to its periodic orbits. For reasons that will be explained in section 4, we regard (X_A, σ_A) as the *Markov chain* defined by A. (Some readers may find it helpful to read now the first four paragraphs of section 4.)

Let B also be a matrix over R_k^+. We will call a block map $\phi : X_A \to X_B$ *weight-preserving* if $\text{wps}_B(\phi(\gamma)) = \text{wps}_A(\gamma)$ for every cycle (periodic orbit) γ of (X_A, σ_A). A *factor map* of (X_A, σ_A) onto (X_B, σ_B) is given by a weight-preserving factor map of the underlying shifts of finite type. We indicate such a map by writing $\phi : (X_A, \sigma_A) \to (X_B, \sigma_B)$, and say that (X_B, σ_B) is a *factor* of (X_A, σ_A) and that (X_A, σ_A) is a *extension* of (X_B, σ_B). In particular, (X_B, σ_B) is a *right-closing factor* of (X_A, σ_A) if there exists a right-closing (weight-preserving) factor map from (X_A, σ_B) to (X_B, σ_B). Also, (X_B, σ_B) is a *right-closing eventual factor* of (X_A, σ_A) if (X_{B^n}, σ_{B^n}) is a right-closing factor of (X_{A^n}, σ_{A^n}) for all large n. A weight-preserving topological conjugacy is called a *block isomorphism*. The following condition is clearly necessary for the existence of a finite-to-one factor map from (X_A, σ_A) to (X_B, σ_B): For each periodic orbit γ of (X_A, σ_A) there exists in (X_B, σ_B) a periodic orbit of the same weight-per-symbol whose (least) period divides that of γ. We refer to this as the *factor periodic point condition* for A, B.

We recall from [**T1, MT1, T3**] some of the objects associated with (X_A, σ_A). Let $\mathbb{R}^{++}$ denote the positive reals. For $x_1, x_2, \ldots, x_k \in \mathbb{R}^{++}$, let $\beta_A(x_1, x_2, \ldots, x_k)$ be the maximum eigenvalue of the nonnegative real-valued matrix $A(x_1, \ldots, x_k)$ provided by the Perron-Frobenius theorem [**S**]. This defines a continuous function $\beta_A : (\mathbb{R}^{++})^k \to \mathbb{R}^{++}$ called the *beta function* of A. Let $W(A)$ denote the convex hull, in $\mathbb{Q}^k$, of

$$\{\text{wps}_A(\gamma) : \gamma \text{ is a cycle of } G(A)\}.$$

Since $G(A)$ has finitely many simple cycles and every cycle can be decomposed into simple ones, $W(A)$ is a polytope, called the *weight-per-symbol polytope* of A. Let F be a nontrivial face of $W(A)$. Define a subgraph of $G(A)$ by retaining an edge e of $G(A)$ (and its weight $\text{wt}_A(e)$) if and only if there exists a cycle γ which contains e and has $\text{wt}_A(\gamma) \in F$. The resulting subgraph is nonwandering; that is, it consists only of a finite number of irreducible components, which we call the *F-components* of A. When $F = W(A)$, we have a single $W(A)$-component, namely, A itself.

These definitions apply to $p \in R_k^+$ by viewing it as a 1×1 matrix and the corresponding objects are quite simple: $\beta_p = p$, the weight-per-symbol polytope of p coincides with the Newton polyhedron and, for each $F \in \mathcal{F}(p)$, there is a single F-component of p, given by the F-face p_F of p. The groups $\Gamma(p)$ and $\Delta(p)$ are generated by $\{x^w : w \in \text{Log}(p)\}$ and $\{x^{u-w} : u, w \in \text{Log}(p)\}$, respectively. We have $c_p \Delta(p) = x^w \Delta(p)$ for any $w \in \text{Log}(p)$. As we shall see in section 4, (X_p, σ_p) represents a *Bernoulli shift*.

We will denote the boundary of a polytope $W \subset \mathbb{Q}^k$ by ∂W. By definition, ∂W is the union of all proper faces of W. Equivalently, ∂W is the topological boundary of W taken in its affine hull in $\mathbb{Q}^k$. Let $v \in \mathbb{Z}^k$. Write $a = \min\{w \cdot v : w \in W\}$. The set $\{w \in W : w \cdot v = a\}$ is then a face of the polytope W; we say that v *exposes* the face $\{w \in W : w \cdot v = a\}$. For $q \in R_k$, writing $\delta_v(q) = \min\{w \cdot v : w \in \mathrm{Log}(q)\}$ and $q = \sum_{w \in \mathrm{Log}(q)} q_w x^w$, we will consider

$$F_v(q) = \{w \in \mathrm{Log}(q) : w \cdot v = \delta_v(q)\}$$

and

$$q_{(v)} = \sum_{w \in F_v(q)} q_w x^w.$$

Fixing $p \in R_k^+$, we assume $\beta_A = p$ since this is necessary [**T1**, **T3**] for the existence of a right-closing (in fact, any finite-to-one) factor map of (X_{A^n}, σ_{A^n}) onto (X_{p^n}, σ_{p^n}) for any n. Let us agree that, unless otherwise indicated, the term *right eigenvector* (of A) will mean a nontrivial vector r over R_k such that $Ar = pr$. Any column of the adjoint $\mathrm{Adj}(pI - A)$ is a right eigenvector for A. Cancelling any common factors the entries of a right eigenvector might have, we obtain a right eigenvector r_A^0 whose entries are coprime. As a result of the uniqueness in the Perron-Frobenius theorem [**S**], r_A^0 is unique up to multiplication by a monomial; we call r_A^0 the *primitive* right eigenvector of A. (Note that monomials are precisely the units of R_k.) The eigenvector r was used in [**T3**] to incorporate the F-components of A in a more general structure, which we now recall. Let $v \in \mathbb{Z}^k$ and let F be the face of $W(p)$ exposed by v. For $I \in S(G(A))$, put

$$r_v(I) = (r(I))_{(v)}$$

to obtain a vector r_v indexed by $S(G(A))$. Define a (weighted) subgraph G of $G(A)$ by letting an edge $e \in E(G(A))$ belong to $E(G)$ (and have weight $\mathrm{wt}_A(e)$) if and only if

$$\delta_v\left(\mathrm{wt}_A(e)\, r(t(e))\right) = \delta_v\left(p\, r(s(e))\right).$$

Let A_v be the R_k^+-matrix with $G(A_v) = G$. We think of A_v and r_v as the v-face of A and r, respectively. It was shown in section 3 of [**T3**] that $S(G(A_v)) = S(G(A))$, no row of A_v is trivial, $A_v r_v = p_F r_v$, and that the irreducible components of A_v are precisely the F-components of A. Furthermore, the beta function of an F-component either equals p_F or it is strictly less than p_F on all of $(\mathbb{R}^{++})^k$, according to whether the component is a sink of A_v or not; the F-components in the former category are said to be *principal*. We will make frequent use of these ideas and facts from [**T3**]; familiarity with section 3 of [**T3**], at least through corollary (9), will be very useful.

The following facts contained in (5.1) and (6.1) of [**MT1**] will also be fundamental for our work: If $\beta_A = p \in R_k^+$ then $W(A) = W(p)$, $\Gamma(A) \subset \Gamma(p)$, $\Delta(A) \subset \Delta(p)$ and $c_A \Delta(A) \subset c_p \Delta(p)$.

Before ending the section, we associate a ring R_p with the polynomial p. Note that for any monomial x^w we have $\Delta(x^w p) = \Delta(p)$, and find $w \in \mathbb{Z}^k$ such that $\widetilde{p} = x^w p \in R(\Delta(p)) = R(\Delta(\widetilde{p}))$. (For example, we can take any w with $-w \in \mathrm{Log}(p)$.) Let

$$R_p = R(\Delta(p)) \begin{bmatrix} 1 \\ \widetilde{p} \end{bmatrix},$$

the ring obtained by adjoining $\frac{1}{\tilde{p}}$ to the group ring $R(\Delta(p))$. Observe that R_p does not depend on the choice of the monomial multiple $\tilde{p}$ of p such that $\tilde{p} \in R(\Delta(p))$.

3
The main results

To introduce the modules that will feature in our results, consider an aperiodic R_k^+-matrix B together with a right eigenvector r. Suppose that $\beta_B = q \in R_k^+$ and that q, B, r are over $R(\Delta(q))$. Using (5.1) of [**PS**] (or (1.9) of [**MT1**]), find a diagonal matrix D such that D has elements of $\Delta(q)$ for its diagonal entries and $\tilde{B} = \frac{1}{c_B} DBD^{-1}$ is over $R(\Delta(B))$. Let $\tilde{r} = Dr$ and $\tilde{q} = q/c_B$. Note that $\tilde{q} \in R(\Delta(q))$ and that

$$R_B = R(\Delta(B)) \begin{bmatrix} 1 \\ \frac{1}{\tilde{q}} \end{bmatrix}$$

is a subring of $R_q = R(\Delta(q)) \begin{bmatrix} \frac{1}{q} \end{bmatrix}$. Clearly, R_B does not depend on our choice of the representative c_B of $c_B \Delta(B)$. View R_q as a module over R_B. Let $\langle r \rangle$ be the R_B-submodule of R_q generated by the entries of $\tilde{r}$. It is easily checked that $\langle r \rangle$ is well-defined up to multiplication by an element of $\Delta(q)$: If c_B and D are replaced by $\bar{c}_B \in c_B \Delta(B)$ and a diagonal matrix $\bar{D}$ such that $\frac{1}{\bar{c}_B} \bar{D} B \bar{D}^{-1}$ is over $R(\Delta(B))$, then $\langle \bar{D}r \rangle = x^w \langle r \rangle$ for some $w \in \Delta(q)$. In particular $\langle r \rangle = R_q$ if and only if $\langle \bar{D}r \rangle = R_q$. This shows that the issue of whether $\langle r \rangle$ equals R_q, which will be our main concern, is independent of the choice of c_B and D.

We start to formulate the conditions for our main results. Fix $p \in R_k^+$. Let F be a proper face of $W(p)$; that is, a face of $W(p)$ other than the empty set and $W(p)$ itself. Note that for $n \in \mathbb{N}$ we have $W(p^n) = nW(p)$ and, as a consequence of the definition of the F-face p_F of p, also $(p^n)_{nF} = (p_F)^n$. A polynomial $q \in R_k$ will be said to F-*conform* to p if there exists $n \in \mathbb{N}$ so that for some $u \in \mathbb{Z}^k$ we have

(∗) $\mathrm{Log}(q) + u \subset \mathrm{Log}(p^n)$ and $(\mathrm{Log}(q) + u) \cap \mathrm{Log}(p_F^n) \neq \phi$.

Note that if this requirement is satisfied for some n then it is satisfied for all larger values of n, though the vector u may vary with n. Let us say that q has a (well-defined) F-*face* if $q_{(v)} = q_{(v')}$ whenever v, v' both expose F; in this case, write q_F for this polynomial. Note that if q F-conforms to p then q has a well-defined F-face; in fact (∗) implies that we have

$$F_v(q) = \mathrm{Log}(q) \cap (\mathrm{Log}(p_F^n) - u)$$

whenever v exposes F. Hence F-conformity of q to p amounts to an affirmative answer to the following question. Is there a well-defined F-face q_F of q corresponding to F and, upon letting the "shape" $\mathrm{Log}(p^n)$ grow, can we apply a translation to simultaneously embed $\mathrm{Log}(q)$ into $\mathrm{Log}(p^n)$ and $\mathrm{Log}(q_F)$ into $\mathrm{Log}(p_F^n)$?

It should be apparent from the above discussion that any v exposing F can be used to check F-conformity: Are there n and u such that

$$\mathrm{Log}(q) + u \subset \mathrm{Log}(p^n) \quad \text{and} \quad \mathrm{Log}(q_{(v)}) + u \subset \mathrm{Log}(p_F^n)\,?$$

The checkability of F-conformity will be further clarified in (3.7). Examples involving F-conformity are given at the end of the section.

To extend our notation to the face $F = W(p)$, we write $q_F = q$ when $F = W(p)$.

Fix an irreducible matrix A over R_k^+ with $\beta_A = p$. Our main results will require the existence of a right eigenvector r of A which satisfies three conditions. The first two are:

(a) The entries of r lie in R_k^+.
(b) For each proper face F of $W(p)$ and each state I that belongs to an F-component of A, the entry $r(I)$ F-conforms to p.

REMARK 3.1. Suppose A is irreducible, $\beta_A = p$, and conditions (a) and (b) hold for a right eigenvector r. For a nontrivial face F of $W(p)$ and a principal F-component $\mathcal{C}$, let $A_\mathcal{C}$ be the matrix representing the principal F-component, so that $A_\mathcal{C}$ is indexed by the states of $\mathcal{C}$ and contains the monomials of A corresponding to the edges in $\mathcal{C}$. It follows from (b) that, whenever I is a state of an F-component $\mathcal{C}$ of A, the entry $r(I)$ has a well-defined F-face $r(I)_F$. Let $r_\mathcal{C}$ denote the vector obtained on taking F-faces $r(I)_F$ of the entries of r corresponding to states I of $\mathcal{C}$. By (5) of [**T3**],

$$A_\mathcal{C}\, r_\mathcal{C} = p_F\, r_\mathcal{C}.$$

As a consequence of (b), we can find $n \in \mathbb{N}$ such that, for any state $I \in \mathcal{C}$, there is $u_I \in \mathbb{Z}^k$ with

$$\mathrm{Log}(r_\mathcal{C}(I)) = \mathrm{Log}(r(I)_F) \subset \mathrm{Log}(p_F^n) - u_I.$$

Index by the states of $\mathcal{C}$ a diagonal matrix D with $D(I,I) = x^{u_I}/c_{p_F}^n$ for $I \in \mathcal{C}$. Let

$$\widehat{A}_\mathcal{C} = \frac{D A_\mathcal{C} D^{-1}}{c_{p_F}},$$

$\widehat{p}_F = p_F/c_{p_F}$ and $\widehat{r}_\mathcal{C} = D r_\mathcal{C}$. Clearly $\widehat{p}_F \in R(\Delta(p_F))$ and $\widehat{r}_\mathcal{C}$ is over $R(\Delta(p_F))$. Since we have

$$\widehat{A}_\mathcal{C}\, \widehat{r}_\mathcal{C} = \widehat{p}_F\, \widehat{r}_\mathcal{C},$$

it follows that $\widehat{A}_\mathcal{C}$ is also over $R(\Delta(p_F))$. These objects will be used to formulate our third condition.

Letting $\mathcal{C}, \widehat{A}_\mathcal{C}, \widehat{p}_F, \widehat{r}_\mathcal{C}$ be as in (3.1) and letting $\widehat{A}_\mathcal{C}$ have period d, consider a period class C of $\mathcal{C}$. Denote by A_C and r_C the restrictions to C of $(\widehat{A}_\mathcal{C})^d$ and $\widehat{r}_\mathcal{C}$, respectively. Note that A_C is aperiodic and $A_C r_C = (\widehat{p}_F)^d r_C$. Moreover, since $\Delta(A_C) = \Delta(A_\mathcal{C})$ and $\Delta((\widehat{p}_F)^d) = \Delta(\widehat{p}_F) = \Delta(p_F)$ by (1.6) of [**MT1**], the objects $(\widehat{p}_F)^d$, A_C, r_C are over $R(\Delta((\widehat{p}_F)^d)) = R(\Delta(p_F))$. So, we are in the situation discussed at the beginning of the section. Consider R_{A_C} and the R_{A_C}-submodule $\langle r_C \rangle$ of $R_{(p_F)^d} = R_{p_F}$ defined at the beginning of the section. Our third requirement is that $\langle r_C \rangle$ equal R_{p_F}:

(c) If F is a nontrivial face of $W(p)$ and C is a period class of a principal F-component of A, then the R_{A_C}-module $\langle r_C \rangle$ equals R_{p_F}.

We are now in a position to state two of our main results.

THEOREM 3.2. *Let A be an irreducible matrix over R_k^+ and let $p \in R_k^+$. Then (X_p, σ_p) is a right-closing eventual factor of (X_A, σ_A) if and only if $\beta_A = p$ and there exists a right eigenvector r of A satisfying (a)–(c).*

THEOREM 3.3. *Let A be an irreducible matrix over R_k^+ and let $p \in R_k^+$. Then (X_p, σ_p) is a right-closing factor of (X_A, σ_A) if and only if $\beta_A = p$, there exists a right eigenvector r of A satisfying (a)–(c), and the factor periodic point condition holds. Moreover, (X_p, σ_p) is a factor of (X_A, σ_A) by a right-closing map of degree 1 if and only if, in addition, A is aperiodic, $\Delta(A) = \Delta(p)$ and $c_A \Delta(A) = c_p \Delta(p)$.*

It was shown in [**MT3**] that any irreducible A with $\beta_A = p$ has a right eigenvector satisfying the positivity condition (a). The global counterpart of (b) is the requirement that, for each state I, a translate of $\text{Log}(r(I))$ be contained in $\text{Log}(p^n)$. As we shall see in (3.4), this follows from what is specified in (b) for proper faces. We shall also indicate in (3.4) that in the case $F = W(p)$ the module condition of (c) is the proper analogue of the dimension group condition of the eventual factors theorem of [**BMT**] (given in [**LM**] as (12.1.4)). With the consideration of other faces, (c) formulates a scaffold of module conditions along the lines of [**MT1**]. Thus (c) is a global module condition together with its (facial) boundary counterparts. It is important to note that (b) goes a step beyond formulating boundary counterparts of the related global condition: For each proper face F and state I of an F-component, (b) not only asks that $\text{Log}(r(I)_F) + u \subset \text{Log}(p_F^n)$ but also makes the near-boundary requirement that we have $\text{Log}(r(I)) + u \subset \text{Log}(p^n)$ for this u. The proofs of (3.2) and (3.3) will take up sections 5–9.

REMARK 3.4. (i) It follows from (b) that, for each state I of A and all large n, a translate of $\text{Log}(r(I))$ is contained in $\text{Log}(p^n)$: By (1.1) of [**KMT**] (or by (3.5) below), it suffices to check that a monomial multiple of $r(I)$ lies in $R(\Delta(p))$. Take $F = W(p)$ in (3.1). Then $A_C = A$, $r_C = r$, $p_F = p$ and $\Delta = \Delta(p)$. Considering any proper face F' of $W(p)$ and taking I_0 to be any state of any principal F'-component of A, it follows from (b) that $\text{Log}(r(I_0))$ is contained in a single coset of $\text{Log}(\Delta)$. Repeat the argument of (3.1) with this I_0 to see that there exists a monomial x^w such that each entry $r(I)$ of r is of the form $x^w q$ with $q \in R(\Delta(p))$.

(ii) Consider $F \in \mathcal{F}(p)$, a principal F-component $\mathcal{C}$ of period d and a period class C of $\mathcal{C}$. Let $\widehat{p}_F, A_C, A_C, r_C$ be as above and let $\widetilde{A}_C, \widetilde{(p_F)^d}, \widetilde{r}_C$ be the objects resulting from taking $B = A_C$, $q = (\widehat{p}_F)^d$, $r = r_C$ at the beginning of this section. Let $\widetilde{p}_F = p_F/c_{p_F}$. From (1.6), (5.5) and (6.1) of [**MT1**] we see that $\Delta(\widetilde{p}_F) = \Delta(p_F) = \Delta((p_F)^d) = \Delta\left(\widetilde{(p_F)^d}\right)$, $\Delta(\widetilde{A}_C) = \Delta(A_C) = \Delta(A_C) \subset \Delta(p_F)$, and that the quotient group $\Delta(p_F)/\Delta(A_C)$ is finite. Since $(\widetilde{p}_F)^d = x^u \widetilde{(p_F)^d}$ for a monomial $x^u \in \Delta(p_F)$, it is easy to see that $\langle r_C \rangle = R_{p_F}$ if and only if we can find $n \in \mathbb{Z}^+$ and, for each element $x^w \Delta(A_C) \in \Delta(p_F)/\Delta(A_C)$, polynomials $q_C^{(w)}(J) \in R(\Delta(A_C))$ such that

$$\sum_{J \in C} q_C^{(w)}(J) \widetilde{r}_C(J) = x^w (\widetilde{p}_F)^n .$$

Clearly, we then obtain similar equations for all larger values of n. (The polynomials $q_C^{(w)}(J) \in R(\Delta(A_C))$ will of course change with n.)

(iii) We quickly indicate the relationship of (c) to the dimension modules of [**T2**]. The dimension module of p is R_p, regarded as a module over $R(\Delta(p))$. When A is aperiodic, its dimension module is over $R(\Delta(A))$. Using the fact that R_p may also be regarded as a $R(\Delta(A))$-module, the equation $Ar = pr$ furnishes a natural $R(\Delta(A))$-module homomorphism of the dimension module of A into R_p. Surjectivity of this homomorphism is equivalent to the requirement $\langle r \rangle = R_p$ made by (c) for the case $F = W(p)$. Thus, the case $F = W(p)$ of (c) is the proper generalization of the dimension group condition of the eventual factors theorem of [**BMT**]. More generally, (c) allows for periodicity, considers maps of dimension modules resulting from the equations $A_C r_C = p_F r_C$ for F-components C of A, and insists that each of these maps be surjective. This paper will not make any explicit use of the relationship of (c) to dimension modules. The action of a right-closing map on scaffolds of dimension modules will be discussed elsewhere.

In advance of further remarks on F-conformity, we recall the description of $\mathrm{Log}(p^n)$ given in [**MT4**]. If $H \geq 0$ and $S \subset \mathbb{Q}^k$, we use the usual norm and distance in $\mathbb{Q}^k$ to write

$$B(S, H) = \{w \in \mathbb{Q}^k : \mathrm{dist}(w, S) < H\}.$$

We put

$$\mathrm{Int}(p^n, H) = (W(p^n) \setminus B(\partial W(p^n), H)) \cap \mathrm{Log}\left(c_p^n \Delta(p)\right) .$$

In other words, $\mathrm{Int}(p^n, H)$ is the set of points of $\mathrm{Log}\left(c_p^n \Delta(p)\right)$ remaining in $W(p^n)$ after we remove a band of width H from around the boundary.

PROPOSITION 3.5. [**MT4**] *Let $p \in R_k^+$. Then*
(i) $p^n \in c_p^n R(\Delta(p))$,
(ii) *there exists $H(p) \geq 0$ such that*

$$\mathrm{Int}(p^n, H(p)) \subset \mathrm{Log}(p^n)$$

for all $n \in \mathbb{N}$.

THEOREM 3.6. [**MT4**] *Let $p \in R_k^+$. Suppose that, for each $F \in \mathcal{F}(p)$, we have constants $K_F \geq H(p_F)$ and $D_F \geq 0$. Then there exist $H_F \geq K_F$, $N \in \mathbb{N}$ and finite subsets $T(F) \subset \mathbb{Z}^k$ such that $T(W(p)) = \{0\}$ and*

$$\mathrm{Log}(p^n) = \bigcup_{F \in \mathcal{F}(p)} (\mathrm{Int}(p_F^n, H_F + D_F) + T(F)) = \bigcup_{F \in \mathcal{F}(p)} (\mathrm{Int}(p_F^n, H_F) + T(F))$$

for all $n \geq N$.

Beyond its use in the following remark, theorem (3.6) gives the structure around which we will construct inductions on faces of $W(p)$ during the proofs of our main results.

REMARK 3.7. Let F be a nontrivial face of $W(p)$, exposed by $v \in \mathbb{Z}^k$, and let $D > 0$. It is not hard to see from (3.6) that there exist $H > 0$ and $N \in \mathbb{N}$ such that the F-conformity of any $q \in R_k$ with the diameter $\operatorname{diam}(\operatorname{Log}(q)) \leq D$ may be checked as follows: Letting $u = b - a$ for any $b \in \operatorname{Int}\left((p_F)^N, H\right)$ and any $a \in \operatorname{Log}(q_{(v)})$, the polynomial q F-conforms to p if and only if
$$\operatorname{Log}(q) + u \subset \operatorname{Log}(p^N).$$
This easily leads to a formulation in terms of $\Delta(p_F)$: There exists a finite set $T \subset \mathbb{Z}^k$ such that F-conformity to p of any $q \in R_k$ with $\operatorname{diam}(\operatorname{Log}(q)) \leq D$ is equivalent to
$$\operatorname{Log}(q) - c_{q_{(v)}} \subset \operatorname{Log}(\Delta(p_F)) + T.$$

We will end the section with examples concerning F-conformity and conditions (a)–(c). In several examples we will have $k = 1$ and write y for the one variable present. A polynomial $p \in R_1^+ = \mathbb{Z}^+[y^\pm]$ has two proper faces, corresponding to $\min(\operatorname{Log}(p))$ and $\max(\operatorname{Log}(p))$, and exposed upon multiplication by 1 and -1. The situation is symmetric; we focus on the face F corresponding to $\min(\operatorname{Log}(p))$. Considering F-conformity of $q \in \mathbb{Z}[y^\pm]$, in the one-variable case we always have a corresponding face q_F, with $\operatorname{Log}(q_F) = \min(\operatorname{Log}(q))$. However, gaps in $\operatorname{Log}(p^n)$ may pose obstructions to F-conformity, a fact exploited in (3.8).

EXAMPLE 3.8. Let $p = 1 + y^2 + y^3$ and $q = 1 + y$. Since $\operatorname{Log}(p^n)$ will not contain 1 for any $n \in \mathbb{N}$, the polynomial q does not F-conform to p. Consider the matrices
$$A = \begin{bmatrix} 1 + y^2 & y^3 + y^4 \\ y^2 & 1 \end{bmatrix} \text{ and } B = \begin{bmatrix} y^2 + y^3 & 1 + y \\ y^2 & 1 \end{bmatrix}.$$
It is easy to see that $\beta_A = \beta_B = p$; in fact the column vector $r^0 = (1 + y, 1)^{tr}$ satisfies both $Ar^0 = pr^0$ and $Br^0 = pr^0$. Note that, as our notation suggests, r^0 has coprime entries. With $S(A) = S(B) = \{1, 2\}$, the corresponding graphs are as follows.

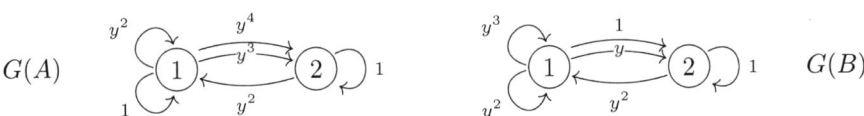

The matrix A has two F-components, given by the subgraphs

and

Note that state 1 belongs to one of the F-components of A and $r^0(1) = 1 + y$ does not F-conform to p. We explain that consequently the Bernoulli shift (X_p, σ_p) is not an eventual right-closing factor of (X_A, σ_A): Any right eigenvector r of A is of the form $r = sr^0$ for some $s \in \mathbb{Z}[y^\pm]$. Let $\delta = \min(\operatorname{Log}(s))$. Then $\delta = \min(\operatorname{Log}(r(1)))$ also.

The only way to get $r(1) = (1+y)s$ to F-conform to p is to force $\delta+1 \notin \text{Log}(r(1))$ by having $\delta+1 \in \text{Log}(s)$ and $s_{\delta+1} = -s_\delta$. But then $r(2) = s$ would have a negative coefficient, violating (a). This shows that no right eigenvector of A satisfies both (a) and (b). It follows from (3.2) that (X_p, σ_p) is not an eventual right-closing factor (and therefore not a right-closing factor) of (X_A, σ_A). On the other hand, despite the entry $r^0(1) = 1+y$, the Markov chain (X_B, σ_B) factors onto (X_p, σ_p) by a right-closing map of degree 1: Using the fact that B has only one F-component which consists of the self-loop at state 2, it is easy to see that r^0 satisfies (b). Indeed, it is easy to check that B and r^0 satisfy the conditions of both (3.2) and (3.3).

EXAMPLE 3.9. In this example we have two variables, x and y. Take $p = 1+x+y$, and let A, B be the $\mathbb{Z}^+[x^\pm, y^\pm]$-matrices whose graphs are given below.

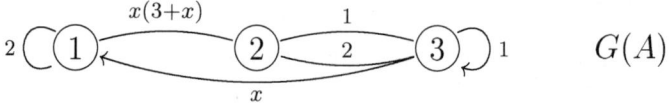

Note that $\beta_A = \beta_B = p$ and that the column vector $r^0 = (x+y, 1)^{tr}$ is the primitive right eigenvector of both A and B. Let F be the face of $W(p)$ consisting of the extreme point $(0,0)$. The polynomial $q = r^0(1) = x + y$ does not F-conform to p. In fact, for no $s \in \mathbb{Z}[x^\pm, y^\pm]$ does sq F-conform to p. One way of seeing this is to observe that sq does not have a well-defined F-face. It may be easiest to use $v = (1,1)$ to expose F as a face of $W(p)$: The face of $W(q)$ exposed by $(1,1)$ is $W(q)$ itself, so that $q_{(v)} = q$ and $(sq)_v = s_{(v)}q_{(v)} = s_{(v)}(x+y)$ will never be a monomial. Since state 1 belongs to an F-component of A, it follows that (b) fails and (X_p, σ_p) is not an (eventual) right-closing factor of (X_A, σ_A). In the case of B state 1 does not lie in an F-component, so non-F-conformity of $r^0(1)$ ceases to be an obstruction. It is easily verified that B and r^0 satisfy all the requirements of (3.3). Thus, (X_B, σ_B) factors onto (X_p, σ_p) by a right-closing map of degree 1.

EXAMPLE 3.10. In this example we have one variable, y. We let $p = 2+y$ and let F be the face of $W(p)$ consisting of the extreme point 0. We let A, B be the $\mathbb{Z}^+[y^\pm]$-matrices whose graphs are given below.

Here, the edge labels indicate the corresponding entries of the matrix. For example, $A(1,2) = y(3+y)$ and the corresponding edge in $G(A)$ represents four edges with weights 1, 1, 1 and y. Note that $\beta_A = \beta_B = p$ and that $r^0 = (3+y, 1, 2+y)^{tr}$ is the primitive right eigenvector of both A and B. Clearly, $p_F = 2$ and $\Delta(p_F) = \{1\}$, the trivial group. The matrix A has two F-components, both principal, given by the subgraphs

$$2 \bigcirc 1 \quad \text{and} \quad 2 \rightleftarrows_2 3 \circlearrowright 1 \quad .$$

Let $\mathcal{C}$ be the first of these. The corresponding matrix $A_{\mathcal{C}}$ equals p_F, which implies that $\Delta(A_{\mathcal{C}}) = \Delta(p_F) = \{1\}$, $R(\{1\}) = \mathbb{Z}$ and $R_{A_{\mathcal{C}}} = R_{p_F} = \mathbb{Z}[\frac{1}{2}]$. Since $r^0(1)_F = (3+y)_F = 3$ and r^0 is primitive, for any right eigenvector r of A the face $r(1)_F$ will be divisible by 3 and generate a proper ideal of $\mathbb{Z}[\frac{1}{2}]$. Thus condition (c) fails for $\mathcal{C}$, and (X_p, σ_p) is not an (eventual) right-closing factor of (X_A, σ_A). Meanwhile B has two F-components

$$1 \bigcirc 1 \quad \text{and} \quad 2 \rightleftarrows_4 3$$

and only the second of these is principal. So, the fact that $r^0(1)_F = 3$ is not an obstruction for B. Since $r^0(2)_F = 1$ and $r^0(3)_F = 2$, each of the two period classes of the unique principal F-component of B satisfies (c). It is easily verified that (3.3) applies to B and r^0, so that (X_p, σ_p) is a factor of (X_B, σ_B) by a right-closing map of degree 1.

EXAMPLE 3.11. This time take $p = 1 + y + y^3$ and let

$$A = \begin{bmatrix} y^2 & 1 + 2y + y^4 \\ 1 & y^3 \end{bmatrix}, \quad r = \begin{bmatrix} 1 + y \\ 1 \end{bmatrix}.$$

Clearly $Ar = pr$, the matrix A is aperiodic,

$$\Delta(A) = \Delta(p) = c_p \Delta(p) = c_A \Delta(A) = \{y^n : n \in \mathbb{Z}\},$$

and r satisfies (a)–(c). In particular, by (3.2), (X_p, σ_p) is a right-closing eventual factor of (X_A, σ_A). However, consideration of the fixed point of (X_A, σ_A) with weight y^2 reveals that the factor periodic point condition fails, showing (X_p, σ_p) is not a factor of (X_A, σ_A).

4

Markov chains and regular isomorphism

We start the section with a description of how to pass from stochastic matrices to polynomial matrices. This rests on ideas and results drawn from [**PT2, T1, T2, PS**]; we refer the reader to section 2 of [**T3**] for a more detailed account. (See [**T4**] for some recent related results.)

Markov chains are traditionally defined by stochastic matrices. Given an irreducible stochastic matrix M, we construct a weighted graph $G(M)$ taking weights in $\mathbb{R}^{++}$: The indexing set of M forms $S(G(M))$ and there is an edge from I to J if and only if $M(I, J) > 0$, in which case the weight of this edge equals $M(I, J)$. Let (X_M, σ_M) denote the shift of finite type defined by $G(M)$, and μ_M the Markov measure defined by M. Then μ_M is a σ_M-invariant Borel measure supported by X_M and (X_M, σ_M, μ_M) is the *Markov chain* defined by M. We will write $\text{wt}_M(e)$, $\text{wt}_M(\gamma)$, $\Gamma(M)$, $\Delta(M)$, $c_M\Delta(M)$ for the objects defined in section 2 for the weighted graph $G(M)$. The *weight-per-symbol* of a $G(M)$-cycle γ of length l is

$$\text{wps}_M(\gamma) = \frac{1}{l}\log(\text{wt}_M(\gamma)) \in \mathbb{R}.$$

For $t \in \mathbb{R}$ we consider the matrix $M^{(t)}$ with

$$M^{(t)}(I, J) = \begin{cases} 0 & \text{if } M(I, J) = 0, \\ M(I, J)^t & \text{if } M(I, J) > 0, \end{cases}$$

and let $\beta_M(t)$ be the spectral radius of $M^{(t)}$. This gives a function $\beta_M : \mathbb{R} \to \mathbb{R}^{++}$ called the *beta function* of M.

Let Q be an irreducible stochastic matrix, also. The Markov chain (X_Q, σ_Q, μ_Q) is said to be an *finite-to-one* (respectively, *right-closing, right-resolving*) *factor* of (X_M, σ_M, μ_M) if there exists a finite-to-one (respectively, right-closing, right-resolving) factor map ϕ of the shift of finite type (X_M, σ_M) onto (X_Q, σ_Q) such that $\mu_M \circ \phi^{-1} = \mu_Q$. Also, (X_Q, σ_Q, μ_Q) is a *right-closing eventual factor* of (X_M, σ_M, μ_M) if (X_Q, σ_Q^n, μ_Q) is a right-closing factor of (X_M, σ_M^n, μ_M) for all large n. By [**JKKMS, PT2**], a finite-to-one factor map ϕ from (X_M, σ_M) to (X_Q, σ_Q) satisfies $\mu_M \circ \phi^{-1} = \mu_Q$ if and only if $\text{wps}_Q(\phi(\gamma)) = \text{wps}_M(\gamma)$ for every $G(M)$-cycle γ. Let $\Gamma \subset \mathbb{R}^{++}$ be a finitely generated subgroup such that $\Gamma(M) \cup \Gamma(Q) \subset \Gamma$, and pick a basis $a_1, a_2, \ldots, a_k$ of the free Abelian group Γ. Upon conjugating M, Q by suitable diagonal matrices over $\mathbb{R}^+$, we obtain (see [**PS, MT1**]) matrices whose nonzero entries lie in Γ. Hence, each nonzero entry of M, Q can be uniquely expressed as a product of integral powers of $a_1, a_2, \ldots, a_k$. Replacing each a_i by an indeterminate x_i, we find that we have obtained from M, Q irreducible matrices A, B with monomials in R_k for their nonzero entries. Clearly, for a finite-to-one factor map ϕ of (X_M, σ_M) onto (X_Q, σ_Q) and a cycle γ of $G(M)$ we have $\text{wps}_Q(\phi(\gamma)) = \text{wps}_M(\gamma)$ if and only if $\text{wps}_B(\phi(\gamma)) = \text{wps}_A(\gamma)$, so that ϕ satisfies $\mu_M \circ \phi^{-1} = \mu_Q$ if and only if it is a (weight-preserving) factor map of (X_A, σ_A) onto (X_B, σ_B).

When the rows of Q are identical (X_Q, σ_Q, μ_Q) is a *Bernoulli shift* and, letting $p = \sum_J B(I, J)$ for any row I, we can identify (X_B, σ_B) with (X_p, σ_p). In other words, Bernoulli shifts correspond to elements of R_k^+, $k \in \mathbb{N}$.

Given a Markov chain (X_M, σ_M, μ_M) and a Bernoulli shift (X_Q, σ_Q, μ_Q) we first check whether $\beta_M = \beta_Q$. If $\beta_M \neq \beta_Q$, the answers to (1), (2) and (3) of the introduction are negative. If $\beta_M = \beta_Q$ then we have $\Gamma(M) \subset \Gamma(Q)$ by (6.1) of [**MT1**], and we take $\Gamma = \Gamma(Q)$ above to consider the matrix A and polynomial p associated above with M and Q. Observing that $\beta_M = \beta_Q$ implies $\beta_A = p$ (see the argument at the bottom of page 159 of [**MT1**]), we obtain from (3.2) and (3.3) the answers to (1) and (2) of the introduction:

THEOREM 4.1. *Let (X_M, σ_M, μ_M) be a Markov chain, and let (X_Q, σ_Q, μ_Q) be a Bernoulli shift. Then (X_Q, σ_Q, μ_Q) is a right-closing eventual factor of (X_M, σ_M, μ_M) if and only if $\beta_M = \beta_Q$ and, considering the R_k^+-matrix A and polynomial $p \in R_k^+$ associated with M, Q, there exists a right eigenvector r of A satisfying (a)–(c) of section 3.*

THEOREM 4.2. *Let (X_M, σ_M, μ_M) be a Markov chain, and let (X_Q, σ_Q, μ_Q) be a Bernoulli shift. Then (X_Q, σ_Q, μ_Q) is a right-closing factor of (X_M, σ_M, μ_M) if and only if $\beta_M = \beta_Q$, and, considering the R_k^+-matrix A and polynomial $p \in R_k^+$ associated with M, Q, there exists a right eigenvector r of A satisfying (a)–(c) of section 3 and the factor periodic point condition holds for A, p. Moreover, (X_Q, σ_Q, μ_Q) is a factor of (X_M, σ_M, μ_M) by a right-closing map of degree 1 if and only if, in addition, M is aperiodic, $\Delta(M) = \Delta(Q)$ and $c_M \Delta(M) = c_Q \Delta(Q)$.*

The *past σ-algebra* $\mathcal{A}_M$ of (X_M, σ_M, μ_M) is generated by the cylinder sets
$$[e_0, e_1, \ldots, e_m]_h = \{x = (x_n) \in X_M : x_h = e_0, x_{h+1} = e_1, \ldots, x_{h+m} = e_m\}$$
such that $h + m \leq 0$. Markov chains (X_M, σ_M, μ_M) and (X_Q, σ_Q, μ_Q) are said to be *regularly isomorphic* if there exists a measure-theoretic isomorphism Φ from (X_M, σ_M, μ_M) to (X_Q, σ_Q, μ_Q) such that $\Phi^{-1}(\mathcal{A}_Q) \subset \sigma_M^{-N}(\mathcal{A}_M)$ and $\Phi(\mathcal{A}_M) \subset \sigma_Q^{-N}(\mathcal{A}_Q)$ for some $N \in \mathbb{Z}^+$. The following answers (3) of the introduction.

THEOREM 4.3. *For a Markov chain (X_M, σ_M, μ_M) and a Bernoulli shift (X_Q, σ_Q, μ_Q), the following three statements are equivalent.*
  (i) *(X_M, σ_M, μ_M) and (X_Q, σ_Q, μ_Q) are regularly isomorphic.*
  (ii) *There exists a Markov chain (X_R, σ_R, μ_R) and right-closing factor maps $\phi : (X_R, \sigma_R, \mu_R) \to (X_M, \sigma_M, \mu_M)$ and $\psi : (X_R, \sigma_R, \mu_R) \to (X_Q, \sigma_Q, \mu_Q)$ of degree 1.*
  (iii) *M is aperiodic, $\beta_M = \beta_Q$, $\Delta(M) = \Delta(Q)$, $c_M \Delta(M) = c_Q \Delta(Q)$ and, considering the R_k^+-matrix A and polynomial $p \in R_k^+$ associated with M, Q, there exists a right eigenvector r of A satisfying (a)–(c) of section 3.*

The proof of (4.3) is preceded by several items that will feed into it.

LEMMA 4.4. *Let A be an irreducible matrix over R_k^+ and let γ be a periodic orbit of (X_A, σ_A). For any $K \in \mathbb{N}$, there exists an aperiodic R_k^+-matrix $\widehat{A}$ and a right-resolving factor map π of $(X_{\widehat{A}}, \sigma_{\widehat{A}})$ onto (X_A, σ_A) such that*

(1) *the pre-image under π of the periodic orbit γ is an orbit of period $K\mathrm{per}(\gamma)$,*
(2) *every periodic point that is not in the orbit γ has a single preimage under π.*

Lemma (4.4) is a version of a lemma of Krieger [**K1**], given in [**LM**] as (10.3.2). To establish (4.4), go through the proof of (10.3.2) in [**LM**] and "lift" weights by assigning to each edge of the extension the weight of its image in $G(A)$.

We denote the affine dimension of a polytope W by $\dim(W)$.

LEMMA 4.5. *Let $p \in R_k^+$. Let $d = \dim(W(p))$ and $m \geq 0$. There exist $N \in \mathbb{N}$ and constants $m_0 \geq m_1 \geq \cdots \geq m_{d-1} = m$ with the following property. If F, G are proper faces of $W(p)$ and we have $w \in B(nF, m_{\dim(F)}) \cap B(nG, m_{\dim(G)})$ for some $n \geq N$, then $F \cap G \neq \emptyset$ and $w \in B(n(F \cap G), m_{\dim(F \cap G)})$.*

PROOF. Let us write m_F for $m_{\dim(F)}$. For each proper face F of $W(p)$, pick $v_F \in \mathbb{Z}^k$ which exposes F. Letting

$$a_F = \min\{W(p) \cdot v_F\} = F \cdot v_F$$

and

$$b_F = \min\{(\mathrm{Log}(p) \setminus F) \cdot v_F\},$$

we have $b_F > a_F$. We let $m_{d-1} = m$ and use induction on the codimension. Suppose, for $i \in \{1, \ldots, d-1\}$, that we have chosen $m_{d-1} \leq \cdots \leq m_i$. Consider proper faces F, G of $W(p)$ with $\dim(F), \dim(G) \geq i$. It is easy to see that if $F \cap G = \emptyset$ then $B(nF, m_F) \cap B(nG, m_G) = \emptyset$ for all large n. Suppose $F \cap G \neq \emptyset$ and $\dim(F \cap G) = i - 1$. Suppose $w \in B(nF, m_F) \cap B(nG, m_G)$. Let $u_1, \ldots, u_h$ be the extreme points of $W(p)$. Note that $u_1, \ldots, u_h \in \mathrm{Log}(p)$. Let w_0 be such that $\|w_0\| < m_i$ and $w - w_0 \in W(p^n)$. Then $w' = w - w_0 \in B(nF, 2m_i) \cap B(nG, 2m_i) \cap W(p^n)$. Express

$$w' = \sum_{j=1}^{h} \alpha_j n u_j,$$

with $\alpha_j \geq 0$ and $\sum_{j=1}^{h} \alpha_j = 1$. Put

$$\alpha'_F = \sum_{u_j \notin F} \alpha_j.$$

Finding $w_F \in nF$ with $\|w' - w_F\| < 2m_i$, we have

$$\begin{aligned}
2m_i \|v_F\| &> (w' - w_F) \cdot v_F \\
&= \sum_{u_j \in F} n\alpha_j a_F + \sum_{u_j \notin F} n\alpha_j (u_j \cdot v_F) - n a_F \\
&= \sum_{u_j \notin F} n\alpha_j (u_j \cdot v_F) - n\alpha'_F a_F \\
&\geq n\alpha'_F b_F - n\alpha'_F a_F.
\end{aligned}$$

Hence,
$$na'_F = \sum_{u_j \notin F} n\alpha_j < 2m_i \|v_F\|/(b_F - a_F).$$

Similarly
$$\sum_{u_j \notin G} n\alpha_j < 2m_i \|v_G\|/(b_G - a_G).$$

Picking j_0 such that $u_{j_0} \in F \cap G$, we have

$$\operatorname{dist}(w, n(F \cap G))$$

$$\leq \left\| w_0 + w' - \sum_{u_j \in F \cap G} n\alpha_j u_j - \left(\sum_{u_j \notin F \cap G} n\alpha_j \right) u_{j_0} \right\|$$

$$\leq m_i + \sum_{u_j \notin F \cap G} n\alpha_j \|u_j - u_{j_0}\|$$

$$\leq m_i + \sum_{u_j \notin F \cap G} n\alpha_j \operatorname{diam}(\operatorname{Log}(p))$$

$$\leq m_i + \operatorname{diam}(\operatorname{Log}(p)) \left(\sum_{u_j \notin F} n\alpha_j + \sum_{u_j \notin G} n\alpha_j \right)$$

$$< m_i + \operatorname{diam}(\operatorname{Log}(p)) \left(\frac{2m_i \|v_F\|}{b_F - a_F} + \frac{2m_i \|v_G\|}{b_G - a_G} \right)$$

We take m_{i-1} to be the maximum of the last quantity over all faces F, G of $W(p)$ with $\dim(F), \dim(G) \geq i$. □

LEMMA 4.6. *Let A be an irreducible matrix over R_k^+. Let F be a proper face of $W(A)$, and let $m \geq 0$. There exists $N \in \mathbb{N}$ such that any $G(A)$-cycle γ with length $l(\gamma) > N$ and $\operatorname{Log}(\operatorname{wt}_A(\gamma)) \in B(l(\gamma)F, m)$ must visit a state of an F-component of A.*

PROOF. Fix a vector $v \in \mathbb{Z}^k$ which exposes F. Put $a = \min\{W(A) \cdot v\} = F \cdot v$. Let b equal the minimum of

$\{\operatorname{wps}_A(\eta) \cdot v : \eta \text{ simple } G(A)\text{-cycle, } \eta \text{ not contained in any } F\text{-component of } A\}.$

Clearly $a < b$. Suppose γ is a $G(A)$-cycle with $\operatorname{Log}(\operatorname{wt}(\gamma)) \in B(l(\gamma)F, m)$. Then

$$\operatorname{Log}(\operatorname{wt}(\gamma)) \cdot v < l(\gamma) a + m\|v\|.$$

Decomposing γ into simple cycles, it is easy to see that if γ does not visit any F-component then

$$\operatorname{Log}(\operatorname{wt}(\gamma)) \cdot v \geq b\, l(\gamma) = a\, l(\gamma) + (b - a)\, l(\gamma).$$

Combining this with the previous inequality, we see that

$$l(\gamma) < m\|v\|/(b - a).$$

So, any γ with $\operatorname{Log}(\operatorname{wt}(\gamma)) \in B(l(\gamma)F, m)$ and length exceeding $m\|v\|/(b-a)$ must visit an F-component of A. □

PROPOSITION 4.7. *Let A be an irreducible matrix over R_k^+. Suppose $\beta_A = p \in R_k^+$ and there is a right eigenvector r of A satisfying (a) and (b). Then there exists $N \in \mathbb{N}$ such that for each periodic orbit γ of (X_A, σ_A) of (least) period exceeding N there exists in (X_p, σ_p) a periodic orbit of the same weight-per-symbol whose period divides that of γ.*

PROOF. Pick $L \in \mathbb{N}$ and, for $F \in \mathcal{F}(p)$ and each state I that lies in an F-component, pick $u_{F,I} \in \mathbb{Z}^k$ such that

$$\mathrm{Log}(r(I)) + u_{F,I} \subset \mathrm{Log}(p^L) \quad \text{and} \quad \mathrm{Log}(r(I)_F) + u_{F,I} \subset \mathrm{Log}(p_F^L).$$

Fix D such that $D \geq \|u_{F,I}\|$ for all $u_{F,I}$ and $D \geq \|u\|$ for all $u \in \mathrm{Log}(p^L) \cup \bigcup_{I \in S(G(A))} \mathrm{Log}(r(I))$. Apply (3.6) to find $H_F \geq H(p_F)$ and finite sets $T(F) \subset \mathbb{Z}^k$ such that, for all large n,

$$(*) \qquad \begin{aligned} \mathrm{Log}(p^n) &= \bigcup_{F \in \mathcal{F}(p)} (\mathrm{Int}(p_F^n, H_F + 4D) + T(F)) \\ &= \bigcup_{F \in \mathcal{F}(p)} (\mathrm{Int}(p_F^n, H_F) + T(F)). \end{aligned}$$

Fix m such that $m \geq H_{W(p)}$ and $m \geq 3D + \max\{\|w\| : w \in T(F), F \in \mathcal{F}(p)\}$. Putting $d = \dim(W(p))$ and $m_{d-1} = m$, choose $m_0 \geq m_1 \geq \cdots \geq m_{d-1}$ as in (4.5). For a proper face F of $W(p)$ write $m_F = m_{\dim(F)}$, and say that a $G(A)$-cycle γ is *near-F* if $\mathrm{Log}(\mathrm{wt}(\gamma)) \in B(l(\gamma)F, m_F)$. Using (3.6) and (4.6), let N be large enough for (*) to hold for all $n \geq N$ and for every near-F cycle of length exceeding N to visit an F-component of A. Also make sure that the conclusion of (4.5) holds for $n \geq N$.

Suppose γ is a $G(A)$-cycle with $l(\gamma) > N$. Put $n = l(\gamma)$. To prove the proposition, we will show that $\mathrm{Log}(\mathrm{wt}(\gamma)) \in \mathrm{Log}(p^n)$. If γ is not near-F for any proper face F then the fact that $m_{d-1} = m \geq H_{W(p)}$ implies that

$$\mathrm{Log}(\mathrm{wt}(\gamma)) \in \mathrm{Int}(p^n, H_{W(p)}) \subset \mathrm{Log}(p^n).$$

To deal with the remaining case, find a proper face F such that γ is near-F but γ is not near-G for any face G with $\dim(G) < \dim(F)$. By (4.6), γ visits a state I of an F-component of A. Hence,

$$\mathrm{Log}(\mathrm{wt}(\gamma)r(I)) \subset \mathrm{Log}(p^n r(I)).$$

Letting $u \in \mathrm{Log}(r(I)_F)$, we have

$$\mathrm{Log}(\mathrm{wt}(\gamma)) + u + u_{F,I} \in \mathrm{Log}(p^n r(I)) + u_{F,I} \subset \mathrm{Log}(p^{n+L}).$$

Using (*), we find $G \in \mathcal{F}(p)$ and $w \in T(G)$ such that

$$\mathrm{Log}(\mathrm{wt}(\gamma)) + u + u_{F,I} \in \mathrm{Int}(p_G^{n+L}, H_G + 4D) + w.$$

Using our choices of D, m and the fact that $m_G \geq m$, it follows that γ is a near-G path. Since γ is also a near-F path, (4.5) shows that γ must, in fact, be near-$(F \cap G)$. Our choice of F then tells us that the face $F \cap G$ cannot have affine dimension lower than $\dim(F)$. So, $F \cap G = F$ and we have $F \subset G$. In particular,

$$u + u_{F,I} \in \mathrm{Log}(p_F^L) \subset \mathrm{Log}(p_G^L) \subset \mathrm{Log}\left(c_{p_G}^L \Delta(p_G)\right).$$

Putting $\widetilde{w} = \mathrm{Log}(\mathrm{wt}(\gamma)) - w$ and $\widetilde{u} = u + u_{F,I}$, we have

$$\widetilde{w} + \widetilde{u} \in \mathrm{Int}(p_G^{n+L}, H_G + 4D).$$

Write $\tilde{w} + \tilde{u} = v + \tilde{v}$ with $v \in \mathrm{Log}(p_G^n)$ and $\tilde{v} \in \mathrm{Log}(p_G^L)$. If $v' \in \partial W(p_G^n)$ then, finding $v'' \in \partial W(p_G^L)$ such that $v' + v'' \in \partial W(p_G^{n+L})$, we see that

$$\begin{aligned} \|v - v'\| &= \|v + \tilde{v} - v' - v'' - \tilde{v} + v''\| \\ &= \|\tilde{w} + \tilde{u} - (v' + v'') - \tilde{v} + v''\| \\ &\geq H_G + 4D - (\|\tilde{v}\| + \|v''\|) \\ &\geq H_G + 2D. \end{aligned}$$

That is,
$$v \in \mathrm{Int}(p_G^n, H_G + 2D).$$

Since $\tilde{v} - \tilde{u} \in \mathrm{Log}(\Delta(p_G))$ and $\|\tilde{v}\|, \|\tilde{u}\| \leq D$, it follows that
$$\tilde{w} = v + \tilde{v} - \tilde{u} \in \mathrm{Int}(p_G^n, H_G)$$
and, considering (*), that
$$\mathrm{Log}(\mathrm{wt}(\gamma)) \in \mathrm{Int}(p_G^n, H_G) + w \subset \mathrm{Log}(p^n). \qquad \square$$

PROOF OF (4.3). The equivalence of (i) and (ii) is a special case of the main result of [**BT**]. Suppose (ii) holds. Then M is aperiodic, $\beta_M = \beta_Q$, $\Delta(M) = \Delta(Q)$ and $c_M \Delta(M) = c_Q \Delta(Q)$ by the results of [**T1, PS**] (see also [**BT**]). To check the rest of (iii), we first use a result of Kitchens [**K**] (see (5.1.11) of [**LM**]) and (13) of [**T1**] to apply a block isomorphism (measure-preserving topological conjugacy) to obtain (X_R, σ_R, μ_R) with a right-resolving factor map $\phi : (X_R, \sigma_R, \mu_R) \to (X_M, \sigma_M, \mu_M)$ and a right-closing one $\psi : (X_R, \sigma_R, \mu_R) \to (X_Q, \sigma_Q, \mu_Q)$. Suppose ϕ is given by the graph homomorphism $(\phi, \phi_0) : G(R) \to G(M)$. Let A, B, p be the R_k^+-matrices and polynomial associated with M, R, Q. Considering the primitive right eigenvector r_A^0 of A, we let $r_B^0(I) = r_A^0(\phi_0(I))$ for $I \in S(G(B))$. Since (ϕ, ϕ_0) is right-resolving, r_B^0 is then the primitive right eigenvector of B. Applying (4.2) to the right-closing factor map ψ, we know that there exists $q \in R_k$ so that qr_B^0 satisfies (a)–(c). Let $r = qr_A^0$. It follows from the definition of r_B^0 that r then satisfies (a). We similarly see that r satisfies (b) and (c), once we recall from (3.13) of [**MT1**] that every F-component of A is the image under (ϕ, ϕ_0) of some F-component of B and that every principal F-component of A is the image of some principal F-component of B. This shows that (ii) implies (iii).

Suppose (iii) holds. Let N be as in (4.7). Let $\gamma_1, \gamma_2, \ldots, \gamma_h$ be a complete list of distinct periodic orbits of (X_A, σ_A) whose periods do not exceed N. Using the fact that $G(p)$ has a single vertex, find edges $e_1, \ldots, e_m$ of $G(p)$ such that $\{\mathrm{Log}(\mathrm{wt}_p(e_j)) : 1 \leq j \leq m\}$ equals the set of extreme points of $W(p)$. Since $W(A) = W(p)$, we can, for $i = 1, 2, \ldots, h$, find $k_j = k_j(i)$, $K = K(i) \in \mathbb{Z}^+$ such that $\sum_{j=1}^m k_j = K$ and
$$\mathrm{wps}_A(\gamma_i) = \sum_{j=1}^m \frac{k_j}{K} \mathrm{Log}(\mathrm{wt}_p(e_j)).$$

For $i = 1, 2, \ldots, h$ let
$$\eta_i = e_1^{k_1} \cdots e_m^{k_m}.$$

Then $l(\eta_i) = K = K(i)$, so that η_i yields a periodic orbit of (X_p, σ_p) whose period divides $K(i)$. Note also that $\text{wps}_p(\eta_i) = \text{wps}_A(\gamma_i)$. Make h applications of (4.4) to find an aperiodic R_k^+-matrix B and a right-resolving factor map $\phi : (X_B, \sigma_B) \to (X_A, \sigma_A)$ of degree 1 such that for $i = 1, 2, \ldots, h$ the inverse image $\phi^{-1}(\gamma_i)$ is a periodic orbit of period $K(i) \text{per}(\gamma_i)$ and every periodic point that is outside $\gamma_1, \gamma_2, \ldots, \gamma_h$ has a single preimage under ϕ. Since ϕ is of degree 1, we have

$$\Delta(B) = \Delta(A) = \Delta(p) \text{ and } c_B \Delta(B) = c_A \Delta(A) = c_p \Delta(p).$$

Working towards an application of (3.3), we next check the factor periodic point condition for B, p. Suppose γ is a periodic orbit of (X_B, σ_B). If $\text{per}(\phi(\gamma)) \leq N$ then $\phi(\gamma) = \gamma_i$ for some $1 \leq i \leq h$. It follows that $\text{per}(\gamma) = K(i) \text{per}(\gamma_i)$ and that η_i is a periodic orbit of (X_p, σ_p) such that $\text{wps}_p(\eta_i) = \text{wps}_B(\gamma)$ and $\text{per}(\eta_i)$ divides $\text{per}(\gamma)$. If $\text{per}(\phi(\gamma)) > N$, we have $\text{per}(\gamma) = \text{per}(\phi(\gamma))$, and we simply appeal to (4.7). To find a right eigenvector r_B of B which fulfills (a)–(c), we consider the graph homomorphism $(\phi, \phi_0) : G(B) \to G(A)$ involved in the right-resolving map ϕ, and define $r_B(I) = r(\phi_0(I))$ for $I \in S(G(B))$. Since the right eigenvector r of A satisfies (a)–(c), it is easy to see, with the aid of (3.13) of [**MT1**], that r_B satisfies (a)–(c). An application of (3.3) now provides a right-closing factor map $\psi : (X_B, \phi_B) \to (X_p, \sigma_p)$ of degree 1. Finally, to return to the setting of stochastic matrices, substitute for $x_1, \ldots, x_k$ the numbers $a_1, \ldots, a_k > 0$ that were used in obtaining A, p. From A, p we immediately recover M and Q, but we may have to conjugate the matrix obtained from B by a diagonal matrix (as in [**PT2, MT1, T3**]) to get a stochastic matrix R. The maps ϕ, ψ are then measure-preserving, since they were constructed to be weight-preserving. This shows (iii) implies (ii), and completes the proof of (4.3). □

5

Necessity of the conditions

We return to the setting of polynomial matrices and fix, for the rest of the paper, an irreducible R_k^+-matrix A with $\beta_A = p \in R_k^+$. In this section, we formulate right-closing maps in terms of state-splittings and use this formulation to deduce (3.3) from (3.2) and prove the necessity of the conditions in (3.2).

We describe the adaptation to our setting of Williams's notion of state-splitting along the lines of [**PT2**]. Section (2.4) of [**LM**] is our standard reference for state-splitting, which originated in [**W**]. Let B be a matrix over R_k^+. A *state-splitting of* (X_B, σ_B) results from a collection of partitions: For every state $I \in S(G(B))$, the set $\{e \in E(G(B)) : s(e) = I\}$ is partitioned into sets $(I,1), (I,2), \ldots, (I, \rho(I))$. For each (I, i), we form a vector $\alpha^{(I,i)}$ indexed by $S(G(B))$ and such that

$$\alpha^{(I,i)}(J) = \sum_{e \in (I,i),\, t(e)=J} \mathrm{wt}_B(e)$$

for each $J \in S(G(B))$. The sum $\sum_{i=1}^{\rho(I)} \alpha^{(I,i)}$ then equals the I-th row of B. This yields a correspondence between partitions of rows of B into vectors over R_k^+ and partitions of the sets $\{e \in S(G(B)) : s(e) = I\}$ and, confusing these two viewpoints, we also speak of *(state-)splitting B*. We use the partition elements (I, i) to index a new matrix $\widehat{B}$ by defining

$$\widehat{B}((I,i), (J,j)) = \alpha^{(I,i)}(J).$$

For each $e \in E(G(B))$ there exists a unique $i \in \{1, 2, \ldots, \rho(s(e))\}$ such that $e \in (s(e), i)$; the graph $G(\widehat{B})$ then contains edges e^j, $1 \leq j \leq \rho(t(e))$, with $\mathrm{wt}_{\widehat{B}}(e^j) = \mathrm{wt}_B(e)$, $s(e^j) = (s(e), i)$ and $t(e^j) = (t(e), j)$. Letting $\psi(e^j) = e$, we obtain a (1-block) left-resolving map $\psi : X_{\widehat{B}} \to X_B$. It is easily checked that ψ is a block isomorphism between $(X_{\widehat{B}}, \sigma_{\widehat{B}})$ and (X_B, σ_B). In addition, if r_B is a right eigenvector of B, we put $r_{\widehat{B}}(I, i) = \alpha^{(I,i)} \cdot r_B$ and calculate:

$$\sum_{(J,j)} \widehat{B}((I,i),(J,j))\, r_{\widehat{B}}(J,j) = \sum_J \alpha^{(I,i)}(J) \sum_j \alpha^{(J,j)} \cdot r_B$$

$$= \sum_J \alpha^{(I,i)}(J) \left(\sum_j \alpha^{(J,j)} \right) \cdot r_B$$

$$= \sum_J \alpha^{(I,i)}(J)\, \beta_B\, r_B(J) = \beta_{\widehat{B}}\, r_{\widehat{B}}(I,i),$$

showing that $r_{\widehat{B}}$ is a right eigenvector of $\widehat{B}$ corresponding to $\beta_{\widehat{B}} = \beta_B$.

When $Ar = pr$ and r is over R_k^+, we will seek state-splittings of a particular form. Writing each entry $r(I)$ as a sum of monomials

$$r(I) = \sum_{i=1}^{\rho(I)} x^{w(I,i)},$$

we will seek to partition the I-th row of A^n into $\rho(I)$ vectors $\alpha^{(I,i)}$, $1 \leq i \leq \rho(I)$, such that

$$\alpha^{(I,i)} \cdot r = x^{w(I,i)} p^n.$$

When we have been able to do this, we say that we have a *splitting of A^n according to r*. The advantage of such a splitting is that the resulting matrix $\widehat{A^n}$ has a monomial right eigenvector: The calculation above shows that the vector whose (I,i) entry equals $\alpha^{(I,i)} \cdot r = x^{w(I,i)} p^n$ is a right eigenvector for $\widehat{A^n}$ and, dividing by the common factor p^n, we obtain a right eigenvector $\widehat{r}$ with $\widehat{r}(I,i) = x^{w(I,i)}$. Hence

$$\sum_{(J,j)} \widehat{A^n}((I,i),(J,j)) \, x^{w(J,j)} = p^n \, x^{w(I,i)},$$

and for each (I,i) we can find a bijection $\Phi^{(I,i)}$ of $\{e \in E(G(\widehat{A^n})) : s(e) = (I,i)\}$ onto $E(G(p^n))$ such that, writing $(J,j) = t(e)$,

$$\mathrm{wt}_{p^n}(\Phi^{(I,i)}(e)) = \mathrm{wt}_{\widehat{A^n}}(e) \, x^{w(J,j)} \, x^{-w(I,i)}.$$

It is easily checked that the maps $\Phi^{(I,i)}$ give a right-resolving factor map $\phi : (X_{\widehat{A^n}}, \sigma_{\widehat{A^n}}) \to (X_{p^n}, \sigma_{p^n})$. Upon composing ϕ with the block isomorphism $\psi^{-1} : (X_{A^n}, \sigma_{A^n}) \to (X_{\widehat{A^n}}, \sigma_{\widehat{A^n}})$, we obtain:

PROPOSITION 5.1. *Let r be a nontrivial vector over R_k^+ such that $Ar = pr$. If A^n can be split according to r then (X_{p^n}, σ_{p^n}) is a right-closing factor of (X_{A^n}, σ_{A^n}).*

We continue:

LEMMA 5.2. *Let r be a nontrivial vector over R_k^+ such that $Ar = pr$, and $N \in \mathbb{N}$. If A^N can be split according to r, so can A^n for all $n \geq N$.*

PROOF. In the above notation, suppose $\alpha^{(I,i)}$ are such that $\alpha^{(I,i)} \cdot r = x^{w(I,i)} p^n$ for each I and $1 \leq i \leq \rho(I)$. Note that the products $\alpha^{(I,i)} A$ then provide a splitting of A^{n+1} according to r, and use induction. □

THEOREM 5.3. *Let A be an irreducible R_k^+-matrix with $\beta_A = p \in R_k^+$. The following are equivalent.*

(i) *(X_{p^n}, σ_{p^n}) is a right-closing factor of (X_{A^n}, σ_{A^n}) for some $n \in \mathbb{N}$.*
(ii) *A has a right eigenvector r over R_k^+ such that, for some $N \in \mathbb{N}$, the matrix A^N can be split according to r.*
(iii) *A has a right eigenvector r over R_k^+ such that A^n can be split according to r for all large n.*
(iv) *(X_p, σ_p) is a right-closing eventual factor of (X_A, σ_A).*

REMARK 5.4. Suppose $B, \bar{B}$ are irreducible R_k^+-matrices with $\beta_B = \beta_{\bar{B}} = q \in R_k^+$ and we have a right-resolving map $\psi : (X_{\bar{B}}, \sigma_{\bar{B}}) \to (X_B, \sigma_B)$ given by the graph homomorphism $(\psi, \psi_0) : G(\bar{B}) \to G(B)$. Let r_B be a right eigenvector of B over R_k^+. We lift r_B to a right eigenvector $r_{\bar{B}} = r_B \circ \psi_0$ of $\bar{B}$ by defining $r_{\bar{B}}(I) = r_B(\psi_0(I))$ for $I \in S(G(\bar{B}))$. It is easy to see from the right-resolving property of ψ that B^l can be split according to r_B if and only if $(\bar{B})^l$ can be split according to $r_B \circ \psi_0$. Note also that r_B equals r_B^0, the primitive right eigenvector of B, if and only if its lift $r_B \circ \psi_0$ equals $r_{\bar{B}}^0$.

PROOF OF (5.3). We know from (5.1) and (5.2) that (ii) implies (iii) and (iii) implies (iv). Since (iv) clearly implies (i), it remains only to prove that (i) implies (ii).

Let $B = A^n$ and let $\phi : (X_B, \sigma_B) \to (X_{p^n}, \sigma_{p^n})$ be a right-closing factor map. We will show that a power B^l may be split according to r. Since each higher block presentation of (X_B, σ_B) factors onto (X_B, σ_B) by a right-resolving map, (5.4) allows us to assume that ϕ is a 1-block map. Next we use a result of [**K**] (given in [**LM**] as (5.1.11)) to find $(X_{\widehat{B}}, \sigma_{\widehat{B}})$ and a block isomorphism $\pi : (X_{\widehat{B}}, \sigma_{\widehat{B}}) \to (X_B, \sigma_B)$ such that $\phi \circ \pi$ is a right-resolving factor map of $(X_{\widehat{B}}, \sigma_{\widehat{B}})$ onto (X_{p^n}, σ_{p^n}). More specifically, $\widehat{B}$ is obtained by partitioning, for some $l \in \mathbb{N}$ and each $I \in S(G(B))$, the set of $G(B)$-paths of length l starting at I into sets (I, i) with the following two properties:

(1) For each (I, i), the paths in (I, i) have the same initial edge, $e^{(I,i)} \in E(G(B))$.
(2) Letting $[I, i] = \{x = (x_n) \in X_B : x_0 x_1 \cdots x_{l-1} \in (I, i)\}$, the sets (I, i) are such that $[I, i] \cap \sigma_B^{-1}[J, j] \neq \phi$ implies

$$\{x \in X_B : x_0 = e^{(I,i)}\} \cap \sigma_B^{-1}[J, j] \subset [I, i].$$

(See [**K, LM**] for the details.) The matrix $\widehat{B}$ is then defined by

$$\widehat{B}((I, i), (J, j)) = \begin{cases} \operatorname{wt}_B(e^{(I,i)}) & \text{if } [I, i] \cap \sigma_B^{-1}[J, j] \neq \phi, \\ 0 & \text{if } [I, i] \cap \sigma_B^{-1}[J, j] = \phi. \end{cases}$$

For each (I, i), let $\alpha^{(I,i)}$ be the vector indexed by $S(G(B)) = S(G(A))$ and with

$$\alpha^{(I,i)}(J) = \sum_{\gamma \in (I,i),\, t(\gamma) = J} \operatorname{wt}_B(\gamma).$$

Then $\sum_i \alpha^{(I,i)}$ equals the I-th row of B^l. Using (2), one checks that the vector $r_{\widehat{B}}$ defined by

$$r_{\widehat{B}}(I, i) = \alpha^{(I,i)} \cdot r_A^0$$

is a right eigenvector of $\widehat{B}$. On the other hand, since $(X_{\widehat{B}}, \sigma_{\widehat{B}})$ factors onto (X_{p^n}, σ_{p^n}) by a right-resolving map, each entry of the primitive right eigenvector $r_{\widehat{B}}^0$ is a monomial. Writing $r_{\widehat{B}}^0(I, i) = x^{w(I,i)}$, there exists $q \in R_k$ such that

(*) $$\alpha^{(I,i)} \cdot r_A^0 = x^{w(I,i)} q$$

for every (I,i). Let $r(I) = \sum_i x^{w(I,i)}$. Summing $(*)$ over i,

$$(**) \qquad r(I)\, q = \left(\sum_i \alpha^{(I,i)}\right)\cdot r_A^0 = p^{nl}\, r_A^0(I)$$

for every $I \in S(G(A))$. Since the entries of r_A^0 are coprime we conclude that $q|p^{nl}$. Let $s \in R_k$ be such that $qs = p^{nl}$. From $(**)$ we see that $r(I) = s r_A^0(I)$ for each I. Multiplying $(*)$ by s, we have

$$\alpha^{(I,i)}\cdot r = x^{w(I,i)}\, p^{nl}$$

for every (I,i). Since $r(I) = \sum_i x^{w(I,i)}$, this shows that the $\alpha^{(I,i)}$ split A^{nl} according to r. □

PROPOSITION 5.5. *Suppose r is a right eigenvector of A over R_k^+ such that for some $n \in \mathbb{N}$ the matrix A^n can be split according to r. Then r satisfies (b) and (c).*

PROOF. Consider $F \in \mathcal{F}(p)$ and an F-component $\mathcal{C}$ of A. Let $A_\mathcal{C}$ be the matrix representing $\mathcal{C}$. Fixing a vector $v \in \mathbb{Z}^k$ which exposes F, we put $r_0(I) = r(I)_{(v)}$ for states I of $\mathcal{C}$. Let d be the period of $\mathcal{C}$. Consider a period class C of $\mathcal{C}$, and fix $I_1 \in C$.

We will establish (b) by showing that $r(J)$ F-conforms to p for every $J \in C$. We will need the following facts from (5) of [**T3**]: For every $I \in C$ and $n \in \mathbb{N}$ there exists $a \in \mathbb{Z}$ so that, for any path γ of $\mathcal{C}$ with $s(\gamma) = I$ and length n, we have

$$(\dagger) \qquad \begin{aligned} \{a\} &= v\cdot \mathrm{Log}(\mathrm{wt}_A(\gamma)\, r_0(t(\gamma))) \\ &= v\cdot \mathrm{Log}(p_F^n r_0(I)) < v\cdot \mathrm{Log}(p^n r(I) - p_F^n r_0(I)), \end{aligned}$$

where the inequality indicates the fact that a is strictly less than every element of the set on the right. Moreover, in the case $\mathcal{C}$ is principal, $(\dagger)$ implies $A_\mathcal{C}\, r_0 = p_F\, r_0$.

Express each $r(I)$ as a sum of monomials,

$$r(I) = \sum_{i=1}^{\rho(I)} x^{w(I,i)},$$

and let (I,i) be a splitting of A^n according to r, so that the vectors $\alpha^{(I,i)}$ associated with (I,i) satisfy $\alpha^{(I,i)}\cdot r = x^{w(I,i)} p^n$. Since the sets (I_1,i), $1 \leq i \leq \rho(I_1)$, partition the $G(A)$-paths of length n starting at I_1, at least one of them contains a path in $G(A_\mathcal{C})$; we order the sets so that $(I_1,1)$ has this property. By replacing $\alpha^{(I,i)}$ by $\alpha^{(I,i)} A^l$ and n by $n+l$, we then assume that n is a multiple of d and that, for each $J \in C$, the set $(I_1,1)$ contains a $G(A_\mathcal{C})$-path γ_J ending at J. We have the equation

$$(\dagger\dagger) \qquad x^{w(I_1,1)}\, p^n = \sum_{\gamma \in (I_1,1)} \mathrm{wt}_A(\gamma)\, r(t(\gamma)).$$

For $J \in C$, let $u = \mathrm{Log}(\mathrm{wt}_A(\gamma_J)) - w(I_1,1)$. It follows from $(\dagger\dagger)$ that $\mathrm{Log}(r(J)) + u \subset \mathrm{Log}(p^n)$. In addition, $(\dagger)$ implies that $w(I_1,1) \in \mathrm{Log}(r_0(I_1))$ and $\mathrm{Log}(r(J)_0) + u \subset \mathrm{Log}(p_F^n)$. This proves (b).

Now suppose $\mathcal{C}$ is principal. For a suitable diagonal matrix D with monomial diagonal entries, we replace A, p, r by DAD^{-1}/c_{p_F}, p/c_{p_F}, Dr to assume, without loss of generality, that $A_\mathcal{C}$ is over $R(\Delta(p_F))$ and $p_F \in R(\Delta(p_F))$. (Here, D is as in

(3.1), except that it has been extended to all of $S(G(A))$ by putting $D(I, I) = 1$ whenever I is not a state of $\mathcal{C}$.) As before, let A_C denote the restriction of $(A_C)^d$ to C and r_C the restriction of r_C to C. Write $c = c_{A_C}$ and $\Delta = \Delta(A_C)$. Let D_0 be a diagonal matrix with monomial diagonal entries such that $\tilde{A}_C = \frac{1}{c} D_0 A_C D_0^{-1}$ is over $\Delta(A_C)$. Putting $\tilde{r}_C = D_0 r_C$, we have the equation

$$\tilde{A}_C \tilde{r}_C = \frac{(p_F)^d}{c} \tilde{r}_C .$$

Recalling that d divides n,

$$(\tilde{A}_C)^{\frac{n}{d}} \tilde{r}_C = \frac{(p_F)^n}{c^{\frac{n}{d}}} \tilde{r}_C .$$

Find $p_{[w]} \in R(\Delta)$ and vectors $\tilde{r}_{[w]}$ over $R(\Delta)$ such that

$$\frac{(p_F)^n}{c^{\frac{n}{d}}} = \sum_{x^w \Delta \in \Delta(p_F)/\Delta} x^w p_{[w]}$$

and

$$\tilde{r}_C = \sum_{x^w \Delta \in \Delta(p_F)/\Delta} x^w \tilde{r}_{[w]} .$$

Using (3.5) (or (1.1) of [**KMT**]) and (5.2), we assume that n is large enough for each $p_{[w]}$ to be nontrivial. Since $\tilde{A}_C$ is over $R(\Delta)$, we have

$$(\tilde{A}_C)^{\frac{n}{d}} \tilde{r}_{[w]} = \sum_{x^u \in \Delta(p_F)/\Delta} p_{[u]} \tilde{r}_{[w-u]} ,$$

from which we see that each $\tilde{r}_{[w]}$ is nontrivial. For $x^w \Delta \in \Delta(p_F)/\Delta$, we can therefore find $I \in C$, $w(I, i) \in \text{Log}(r_C(I))$ and $x^u \in \Delta = \Delta(A_C)$ such that $x^{w(I,i)} D_0(I, I) = x^w x^u$. For $J \in C$ we let

$$q(J) = x^{-u} \sum_{\gamma} \text{wt}_{\tilde{A}_C}(\gamma) ,$$

the sum being over all $G(A_C)$-paths γ such that $\gamma \in (I, i)$ and $t(\gamma) = J$. Then $q(J) \in R(\Delta(A_C))$ and, as a result of the equation

$$x^{w(I,i)} p^n = \sum_{\gamma \in (I,i)} \text{wt}_A(\gamma) \, r(t(\gamma))$$

and (†), we have

$$x^w (p_F)^n / c^{\frac{n}{d}} = \sum_{J \in C} q(J) \tilde{r}_C(J) .$$

This establishes (c). □

We have already remarked on the necessity of $\beta_A = p$ for (3.2). Now (5.3) and (5.5) establish the necessity of the remaining conditions of (3.2), namely the existence a right eigenvector satisfying (a)–(c). We end the section by using [**A1, A2**] to deduce (3.3) from (3.2).

PROOF OF (3.3). We already know the necessity of the given conditions for the existence of a right-closing factor map from (X_A, σ_A) to (X_p, σ_p). Conversely, by (3.2) and (5.3), these conditions provide us with splittings of A^n according to r for all large n. In [**A2**] Ashley shows how to use analogous splittings and the related factor periodic point condition to produce right-closing maps (of the first powers) in the case of shifts of finite type. A right-closing factor map of (X_A, σ_A) onto (X_p, σ_p) is constructed by going through the argument of [**A2**] with our factor periodic point condition and splittings of A^n. Regarding the degree 1 statement, aperiodicity of A is well-known to be necessary, and $\Delta(A) = \Delta(p)$, $c_A \Delta(A) = c_p \Delta(p)$ are necessary by [**PS**]. According to [**A1**], these conditions are also sufficient to replace a right-closing factor map from (X_A, σ_A) to (X_p, σ_p) by a right-closing factor map of degree one. □

6
Totally conforming eigenvectors and the one-variable case

The remainder of the paper is devoted to the sufficiency of the conditions of (3.2). Henceforth we take A to be an irreducible R_k^+-matrix with $\beta_A = p \in R_k^+$ and r to be a right eigenvector of A satisfying (a)–(c). Write $S = S(G(A))$. For $I \in S$, we refer to the entry $r(I)$ as the *(state) mass* of I and define the *mass* of a $G(A)$-path γ to be

$$\mathrm{mass}(\gamma) = \mathrm{mass}_A(\gamma) = \mathrm{wt}_A(\gamma) r(t(\gamma)).$$

We also write $W_n(I) = \{G(A)\text{-paths } \gamma : s(\gamma) = I,\ l(\gamma) = n\}$.

We will say that the right eigenvector r *totally conforms* to p if, in addition to satisfying (a)–(c), it has the property that every entry $r(I)$ F-conforms to p for every proper face F of $W(p)$.

We will prove that A^n can, for large n, be split according to r. Our proof will consist of two steps, each involving a round of state-splitting. First we will state-split a suitable power A^N to obtain a totally conforming right eigenvector. Then we will show that, when a matrix has a totally conforming right eigenvector, large enough powers of the matrix can be split according to that eigenvector. In the general case each of the state-splittings will require an induction on the faces of $W(p)$.

In this section and the next we focus on the one-variable case, obtaining a totally conforming right eigenvector in this section, and splitting according to this vector in section 7. In dealing with the one-variable case, we are able to present the state-splittings without the additional complication of the inductions on the faces. The general case is then considered in sections 8–9.

So, assume for the rest of this section and section 7 that $k = 1$ and $R = R_1 = \mathbb{Z}[y^{\pm}]$. Write

$$p = p(y) = \sum_{i=0}^{d} p_{w_i} y^{w_i},$$

with d, w_i, $p_{w_i} \in \mathbb{Z}$, $d > 0$, $w_0 < w_1 < \cdots < w_d$ and $p_{w_i} > 0$. In particular, $\mathrm{Log}(p) = \{w_0, w_1, \ldots, w_d\}$ and $W(p)$ has two proper faces $F_0 = \{w_0\}$ and $F_\infty = \{w_d\}$. The face F_0 of $W(p)$ is exposed upon multiplication by 1; we denote the corresponding face of A by A_{F_0}. For $q \in R$, put $\delta_0(q) = \min(\mathrm{Log}(q))$, $\delta_\infty(q) = \max(\mathrm{Log}(q))$. Write $\delta_0(I) = \delta_0(r(I))$ and $\delta_\infty(I) = \delta_\infty(r(I))$. Property (b) of r

states that, for large n, we have

$$\mathrm{Log}(r(I)) - \delta_0(I) + n\delta_0(p) \subset \mathrm{Log}(p^n)$$

when I is a state of an F_0-component of A and

$$\mathrm{Log}(r(I)) - \delta_\infty(I) + n\delta_\infty(p) \subset \mathrm{Log}(p^n)$$

when I is a state of an F_∞-component. Property (c) consists of three sets of module conditions corresponding to the three faces $W(p)$, F_0, F_∞ of $W(p)$. In the case of F_0 and a period class C of a principal F_0-component of period d, we have $\Delta(A_C) = \{1\}$, so that $R_{A_C} = \mathbb{Z}\left[\frac{1}{p_0}\right] = R_{p_0}$ and (c) stipulates that the $\mathbb{Z}\left[\frac{1}{p_0}\right]$-ideal $\langle r(I)_{\delta_0(I)} : I \in C \rangle$ equal $\mathbb{Z}\left[\frac{1}{p_0}\right]$.

For the duration of this section we will divide A and p by y^{w_0} to assume that $\delta_0(p) = w_0 = 0$, and we will write $\delta = \delta_\infty(p)$. Then $p_{F_0} = p_0$. The fact that $0 \in \mathrm{Log}(p)$ implies $\Gamma(p) = \Delta(p)$ and, by (6.1) of [**MT1**], we have $\Gamma(A) \subset \Gamma(p) = \Delta(p)$. Using (1.9) of [**MT1**], we take A to be over $\Gamma(A)$. We also assume that $\mathrm{Log}(p)$ generates $\mathbb{Z}$, so that $R(\Delta(p)) = R$. In this case (3.5) and (3.6) reduce to (6.1), for which we provide a short proof. For $E \subset \mathbb{R}$, we put $E_\mathbb{Z} = E \cap \mathbb{Z}$.

PROPOSITION 6.1. *There exist* $H \in \mathbb{N}$ *and subsets* $E_0 \subset [0, H]_\mathbb{Z}$ *and* $E_\infty \subset [-H, 0]_\mathbb{Z}$ *such that*

$$\mathrm{Log}(p^n) = E_0 \cup [H, n\delta - H]_\mathbb{Z} \cup (E_\infty + n\delta)$$

for all large n.

PROOF. Since $\mathrm{Log}(p) = \{w_0 = 0, w_1, \ldots, w_{d-1}, w_d = \delta\}$, we have

$$\mathrm{Log}(p^n) = \{\sum_{i=0}^{d} n_i w_i : n_i \in \mathbb{Z}^+ \text{ and } \sum_{i=0}^{d} n_i = n\},$$

and the minimum and maximum of $\mathrm{Log}(p^n)$ equal 0 and $n\delta$ respectively. Since $\mathrm{Log}(p)$ generates $\mathbb{Z}$, we can find $l_i \in \mathbb{Z}$ such that $\sum_{i=0}^{d} l_i w_i = 1$. Moreover, since $w_0 = 0$, we can choose l_0 so that $\sum_{i=0}^{d} l_i = 0$. Then

$$\sum_{i=0}^{d} \max\{l_i, 0\} = -\sum_{i=0}^{d} \min\{l_i, 0\} = L$$

for some $L \in \mathbb{N}$. Let $b = -\sum_{i=0}^{d} \min\{l_i, 0\} w_i$. Then $b + 1 = \sum_{i=0}^{d} \max\{l_i, 0\} w_i$. Take $N = \delta L$. □

LEMMA 6.2. *For* $n \geq N$, *we have*

$$[b\delta, (b+1)\delta + (n - \delta L)\delta]_\mathbb{Z} \subset \mathrm{Log}(p^n).$$

PROOF. Let $j \in \{0, 1, \ldots, n - \delta L\}$. Define $n_d = \max\{l_d, 0\}\delta + j$, $n_i = \max\{l_i, 0\}\delta$ for $i = 1, 2, \ldots, d-1$ and $n_0 = n - \sum_{i=1}^{d} n_i$. Since $n_i \geq 0$, $\sum_{i=0}^{d} n_i = n$ and $\sum_{i=0}^{d} n_i w_i = (b+1)\delta + j\delta \in \mathrm{Log}(p^n)$, this shows that every δ-th element of

$[(b+1)\delta, (b+1)\delta + (n - \delta L)\delta]_\mathbb{Z}$ belongs to $\text{Log}(p^n)$. Moreover the n_i are such that if, for $h \in \{0, 1, \ldots, \delta\}$, we let $n'_i = n_i - hl_i$ then $n'_i \geq 0$, $\sum_{i=0}^{d} n'_i = n$ and

$$\sum_{i=0}^{d} n'_i w_i = \sum_{i=0}^{d} n_i w_i - h = (b+1)\delta + j\delta - h.$$

As j and h are arbitrary elements of $\{0, 1, \ldots, n-\delta L\}$ and $\{0, 1, \ldots, \delta\}$ respectively, this proves the lemma. $\square$

Letting $H = \max\{b\delta, (\delta L - b - 1)\delta\}$, we see from (6.2) that $[H, n\delta - H]_\mathbb{Z} \subset \text{Log}(p^n)$ for $n \geq N$. For $n \geq N$, consider the intervals $[0, H]$ and $[n\delta - H, n\delta]$. It is easy to check that an element of $[0, H]_\mathbb{Z}$ belongs to $\text{Log}(p^n)$ if and only if it is of the form $a_1 + a_2 + \cdots + a_H$, with $a_1, \ldots, a_H \in \text{Log}(p)$. So we let

$$E_0 = [0, H] \cap \{a_1 + \cdots + a_H : a_1, \ldots, a_H \in \text{Log}(p)\}.$$

Similarly, with

$$E_\infty = [-H, 0] \cap \{a_1 + \cdots + a_H - H\delta : a_1, \ldots, a_H \in \text{Log}(p)\},$$

an element of $[n\delta - H, n\delta]_\mathbb{Z}$ belongs to $\text{Log}(p^n)$ if and only if it lies in the translate $E_\infty + n\delta$. This completes the proof of (6.1). $\square$

Put

$$M = \max\{r(I)_w : I \in S, w \in \text{Log}(r(I))\}$$

and

$$D = \max\{\delta_\infty(I) - \delta_0(J) : I, J \in S\} = \text{diam}\left(\bigcup_{I \in S} \text{Log}(r(I))\right).$$

Let S_0 denote the union of the states of the F_0-components of A. Let H, E_0, E_∞ be as in (6.1). Note that, since $0 \in \text{Log}(p)$, it follows from (6.1) that $\text{Log}(p^L) \subset E_0 \cup [H, L\delta - H]_\mathbb{Z} \cup (E_\infty + L\delta)$ for every $L \in \mathbb{N}$. (We take $[H, L\delta - H]_\mathbb{Z}$ to be the empty set when $L\delta - H < H$.) Use property (b) of r to find $L \in \mathbb{N}$ such that

$$\text{Log}(r(J)) - \delta_0(J) \subset \text{Log}(p^L)$$

whenever $J \in S_0$. By increasing L if necessary, we assume that $L > (H+D+1)|S|$.

The following proposition is the one-variable case of a result of [**MT4**] which will be stated in full generality in section 9 as (9.1).

PROPOSITION 6.3. *Let $\mathcal{E}_0 \subset \mathbb{Z}^+$ be a finite set such that $\mathcal{E}_0 \cap [0, H] \subset E_0$. Let $K > 0$. There exist $\epsilon > 0$ and $N \in \mathbb{N}$ such that for $a \in \mathbb{Z}^+$ and nonzero $b \in \mathcal{E}_0$ the inequality*

$$(p^n)_{a+b} \geq K (p^n)_a$$

holds whenever $n \geq N$ and $a \leq \epsilon n$.

In this section our use of (6.3) will be confined to the proof of (6.5) below; we will make more extensive use of (6.3) in section 7. Clearly, there is an analogue of (6.3) for $n\delta - \epsilon n \leq u \leq n\delta$, as well as analogues of (6.4) and (6.5) near the high end $n\delta$ of p^n. (Make the change of variable $y \mapsto y^{-1}$.)

As before, we write the state mass $r(I)$ as a sum of monomials:
$$r(I) = \sum_{i=1}^{\rho(I)} y^{w(I,i)}.$$
We will state-split large powers A^n of A by partitioning each $W_n(I)$ into sets (I,i), $1 \leq i \leq \rho(I)$. In this initial step of our proof, the sets (I,i) will not give a splitting of A^n according to r, but they will be such that the associated vectors $\alpha^{(I,i)}$ satisfy the following weaker requirements:
$$\mathrm{Log}(\alpha^{(I,i)} \cdot r) \subset \mathrm{Log}(y^{w(I,i)} p^n),$$
$$\delta_0(\alpha^{(I,i)} \cdot r) = w(I,i),$$
$$\delta_\infty(\alpha^{(I,i)} \cdot r) = w(I,i) + n\delta.$$
Considering the matrix $\widehat{A^n}$ and the right eigenvector $\widehat{r}$ resulting from such a splitting, we have
$$\widehat{r}(I,i) = \alpha^{(I,i)} \cdot r = \sum_{\gamma \in (I,i)} \mathrm{mass}(\gamma),$$
and the above requirements clearly imply that every entry $\widehat{r}(I,i)$ is F-conforming for every proper face F of p^n.

We start to describe n and the state-splitting. Fix I. Each time we assign a path $\gamma \in W_n(I)$ to an atom (I,i) of the partition, we need to have
$$\mathrm{Log}(\mathrm{mass}(\gamma)) \subset \mathrm{Log}(y^{w(I,i)} p^n).$$
According to (6.1), for large n the gaps of $\mathrm{Log}(y^{w(I,i)} p^n)$ occur within distance H of $w(I,i)$ or $w(I,i) + n\delta$. Hence, we have to be careful when $\mathrm{mass}(\gamma)$ is within H of one of the endpoints. Call $\gamma \in W_n(I)$ a *near-F_0 path* if $\mathrm{Log}(\mathrm{mass}(\gamma))$ intersects some $\mathrm{Log}(y^{w(I,i)} p^n)$ within H of $w(I,i)$; that is, γ is a near-F_0 path if and only if
$$\delta_0(\mathrm{mass}(\gamma)) \leq \delta_\infty(I) + H.$$
Similarly, call γ a *near-F_∞ path* if and only if
$$\delta_\infty(\mathrm{mass}(\gamma)) \geq \delta_0(I) + n\delta - H.$$
In order to be sure that the set of near-F_0 paths is disjoint from the set of near-F_∞ paths, it is sufficient to have
$$\delta_\infty(I) + H + D < \delta_0(I) + n\delta - H.$$
So, we insist that $n\delta > 2H + 2D$. We also require $n > L$. Since $L > (H+D+1)|S|$, we will then be able to apply the following lemma.

LEMMA 6.4. *Any near-F_0 path of length greater than $(H+D+1)|S|$ must visit S_0.*

PROOF. If γ is a path whose length exceeds $(H+D+1)|S|$ then γ contains at least $H+D+1$ simple cycles. If none of these simple cycles passes through S_0, then $\mathrm{Log}(\mathrm{wt}(\gamma)) \geq H+D+1$. But for any near-$F_0$ path we have
$$\delta_0(t(\gamma)) + \mathrm{Log}(\mathrm{wt}(\gamma)) = \delta_0(\mathrm{mass}(\gamma)) \leq \delta_\infty(s(\gamma)) + H,$$

which implies $\mathrm{Log}(\mathrm{wt}(\gamma)) \leq H + D$. □

We will ensure $\delta_0(\alpha^{(I,i)} \cdot r) = w(I,i)$ by making $\rho(I)$ initial assignments: For $1 \leq i \leq \rho(I)$, we will assign to the set (I,i) a path $\gamma^{(I,i)}$ with $t(\gamma^{(I,i)}) \in S_0$ and $\delta_0(\mathrm{mass}(\gamma^{(I,i)})) = w(I,i)$. We now show that this is possible. For $w \in \mathrm{Log}(r(I))$, observe that either $(p^n r(I))_w \to \infty$ as $n \to \infty$, or the following applies: $p_0 = 1$ and w is such that
$$(p^l\, r(I))_w = p_0^l\, r(I)_w = r(I)_w$$
for all l.

LEMMA 6.5. *Suppose $w \in \mathrm{Log}(r(I))$ is such that $(p^n r(I))_w \to \infty$ as $n \to \infty$. Then*
$$|\{\gamma \in W_n(I) : t(\gamma) \in S_0,\ \delta_0(\mathrm{mass}(\gamma)) = w\}| \longrightarrow \infty$$
as $n \to \infty$.

PROOF. Let
$$m = \min\{w - \delta_0(\mathrm{mass}(\gamma)) : s(\gamma) = I,\ w \in \mathrm{Log}(\mathrm{mass}(\gamma))\},$$
and let γ be a path such that $s(\gamma) = I$, $w \in \mathrm{Log}(\mathrm{mass}(\gamma))$ and $w - \delta_0(\mathrm{mass}(\gamma)) = m$. Write $J = t(\gamma)$. Observing that $w - \mathrm{Log}(\mathrm{wt}(\gamma)) \in \mathrm{Log}(r(J))$, consider $\eta \in W_n(J)$ with $w - \mathrm{Log}(\mathrm{wt}(\gamma)) \in \mathrm{Log}(\mathrm{mass}(\eta))$. Recall that $\delta_0(\mathrm{mass}(\eta)) \geq \delta_0(J)$, with equality if and only if η is an A_{F_0}-path. Since $w \in \mathrm{Log}(\mathrm{mass}(\gamma\eta))$ and
$$\begin{aligned} w - \delta_0(\mathrm{mass}(\gamma\eta)) &= w - \delta_0(\mathrm{mass}(\gamma)) + \delta_0(J) - \delta_0(\mathrm{mass}(\eta)) \\ &= m + \delta_0(J) - \delta_0(\mathrm{mass}(\eta)), \end{aligned}$$
the minimality of m implies that η is an A_{F_0}-path. This shows that, by extending γ by a suitable A_{F_0}-path, we may assume $J \in S_0$. Assuming $J \in S_0$, consider
$$a = w - \mathrm{Log}(\mathrm{wt}(\gamma)) \in \mathrm{Log}(r(J)).$$
By the above argument, any $\eta \in W_n(J)$ with $a \in \mathrm{Log}(\mathrm{mass}(\eta))$ is an A_{F_0}-path. Hence,
$$(p^n r(J))_a \leq M\, (p^n r(J))_{\delta_0(J)}.$$
Since $(p^n)_{a-\delta_0(J)}\, r(J)_{\delta_0(J)} \leq (p^n r(J))_a$ and $(p^n r(J))_{\delta_0(J)} = (p^n)_0\, r(J)_{\delta_0(J)}$, we then have
$$(p^n)_{a-\delta_0(J)} \leq M\, (p^n)_0.$$
As $a - \delta_0(J) \in \mathrm{Log}(p^L)$, this would contradict (6.3) unless $a = \delta_0(J)$. Therefore $m = 0$. Moreover, for any A_{F_0}-path $\eta \in W_n(J)$, the path $\gamma\eta$ has $\delta_0(\mathrm{mass}(\gamma\eta)) = w$. Consideration of those η that terminate in a principal F_0-component proves the lemma in the case $p_0 > 1$.

Suppose $p_0 = 1$. In this case, each F_0-component consists of a single periodic orbit. Since $(p^n r(I))_w \to \infty$, we can find a path η such that $s(\eta) = I$, $w \in \mathrm{Log}(\mathrm{mass}(\eta))$, $l(\eta) \geq L$ and, writing $\eta = \eta_1 \eta_2$ with $l(\eta_1) = L$, the tail η_2 is not an A_{F_0}-path. Since η_1 has $\delta_0(\mathrm{mass}(\eta_1)) \leq w$, lemma (6.4) tells us that η_1 must visit S_0. Let $I' = t(\eta)$ and $w' = w - \text{Log}(\text{wt}(\eta)) \in \text{Log}(r(I'))$ and apply the above argument to I' and w' to extend η to a path α with the following properties: $s(\alpha) = I$, $w = \delta_0(\text{mass}(\alpha))$, $J = t(\alpha) \in S_0$ and α can be decomposed $\alpha = \alpha_1 \alpha_2$ so that $J_1 = t(\alpha_1) = s(\alpha_2) \in S_0$ and α_2 is not an A_{F_0}-path. Let $\tilde{\alpha}_1$, $\tilde{\alpha}_2$ be A_{F_0}-cycles of the same length l such that $s(\tilde{\alpha}_1) = t(\tilde{\alpha}_1) = J_1$ and $s(\tilde{\alpha}_2) = t(\tilde{\alpha}_2) = J$. For $j = 0, 1, \ldots, n$ the paths

$$\alpha_1 \, (\tilde{\alpha}_1)^j \, \alpha_2 \, (\tilde{\alpha}_2)^{n-j}$$

provide $n+1$ distinct elements of

$$\{\gamma \in W_{l(\alpha)+nl}(I) : t(\gamma) \in S_0, \delta_0(\text{mass}(\gamma)) = w\}. \qquad \square$$

Let W_0 be the set of all paths γ_0 such that $l(\gamma_0) \leq L$, $s(\gamma_0) = I$, $t(\gamma_0) \in S_0$ and γ_0 does not visit S_0 prior to its terminal state. Using (6.4) and considering first entry into S_0, we see that every near-F_0 path γ is of the form $\gamma = \gamma_0 \gamma_1$ for some $\gamma_0 \in W_0$. (Note that, in the case $I \in S_0$, the set W_0 consists of the empty path, so that γ_0 is the empty path and $\gamma_1 = \gamma$.)

LEMMA 6.6. *Suppose $p_0 = 1$ and $w \in \text{Log}(r(I))$ is such that*

(†) $$(p^l \, r(I))_w = p_0^l \, r(I)_w = r(I)_w$$

for all l. Then

$$|\{\gamma \in W_n(I) : t(\gamma) \in S_0, \delta_0(\text{mass}(\gamma)) = w\}| = r(I)_w \, .$$

PROOF. In this case $R_{p_0} = \mathbb{Z}$, each F_0-component of A consists of a single periodic orbit and, by (c), we have $r(J)_{\delta_0(J)} = 1$ for every $J \in S_0$. Letting W_0 be as above and decomposing each near-F_0 path $\gamma = \gamma_0 \gamma_1$ as above, we find that

$$r(I)_w = (p^n r(I))_w = \sum_{\gamma_0 \in W_0} \left(\text{wt}(\gamma_0) \, p^{n-l(\gamma_0)} \, r(t(\gamma_0)) \right)_w .$$

For $\gamma_0 \in W_0$ we have

(††) $$p^{l(\gamma_0)} r(I) - \text{wt}(\gamma_0) r(t(\gamma_0)) \in R^+ .$$

In particular, there are $a \in \text{Log}(r(I))$ and $b \in \text{Log}(p^{l(\gamma_0)})$ such that

$$\text{Log}(\text{wt}(\gamma_0)) + \delta_0(t(\gamma_0)) = a + b \, .$$

Also note that (††) and (†) together show

$$\left(\text{wt}(\gamma_0) \, p^{n-l(\gamma_0)} \, r(t(\gamma_0)) \right)_w = p_0^{n-l(\gamma_0)} \left(\text{wt}(\gamma_0) \, r(t(\gamma_0)) \right)_w = (\text{wt}(\gamma_0) \, r(t(\gamma_0)))_w \, .$$

Suppose $(\text{wt}(\gamma_0) \, r(t(\gamma_0)))_w \neq 0$. Then

$$w - \text{Log}(\text{wt}(\gamma_0)) \in \text{Log}(r(t(\gamma_0))) \, ,$$

which means that there exists $c \in \text{Log}(r(t(\gamma_0)))$ such that $w = a + b + c - \delta_0(t(\gamma_0))$. Recalling

$$\text{Log}(r(t(\gamma_0))) - \delta_0(t(\gamma_0)) \subset \text{Log}(p^L)$$

and applying (†) with $l = l(\gamma_0) + L$, we find that $b = 0 = c - \delta_0(t(\gamma_0))$ and $a = w$. This shows that $\left(\mathrm{wt}(\gamma_0)\, p^{n-l(\gamma_0)}\, r(t(\gamma_0))\right)_w \neq 0$ if and only if $\mathrm{Log}(\mathrm{wt}(\gamma_0)) + \delta_0(t(\gamma_0)) = w$ and, in this case,

$$\left(\mathrm{wt}(\gamma_0)\, p^{n-l(\gamma_0)}\, r(t(\gamma_0))\right)_w = (\mathrm{wt}(\gamma_0)\, r(t(\gamma_0)))_w = r(t(\gamma_0))_{\delta_0(t(\gamma_0))} = 1$$

as a result of (c). Since $t(\gamma_0) \in S_0$ and each F_0-component is a a single orbit, the lemma follows. □

Whenever $w = w(I,i)$ is such that $(p^n r(I))_w \to \infty$ as $n \to \infty$, the existence of the required paths $\gamma^{(I,i)}$ is guaranteed by (6.5) for all large n. The condition $(p^n r(I))_w \to \infty$ fails only when (6.6) applies to provide exactly $r(I)_w$ paths $\gamma \in W_n(I)$ such that $t(\gamma) \in S_0$ and $\delta_0(\mathrm{mass}(\gamma)) = w$; we bijectively assign these paths to the $r(I)_w$ sets (I,i) with $w(I,i) = w$. We have thus assigned to each (I,i) a path $\gamma^{(I,i)} \in W_n(I)$ with $\delta_0(\mathrm{mass}(\gamma^{(I,i)})) = w(I,i)$ and $\mathrm{Log}(\mathrm{mass}(\gamma^{(I,i)})) \subset \mathrm{Log}(y^{w(I,i)} p^n)$. We proceed to the allocation of the remaining near-F_0 paths.

Let $\gamma \notin \{\gamma^{(I,i)} : 1 \leq i \leq \rho(I)\}$ be a near-F_0 path and, as before, write $\gamma = \gamma_0 \gamma_1$ with $\gamma_0 \in W_0$. Put $l = l(\gamma_0)$ and $J = t(\gamma_0)$. The equation $A^l r = p^l r$ implies that

$$\mathrm{Log}(\mathrm{wt}(\gamma_0)) + \delta_0(J) \in \mathrm{Log}(p^l r(I)).$$

Hence, we can find $1 \leq i_0 \leq \rho(I)$ and $w \in \mathrm{Log}(p^l)$ such that

$$\mathrm{Log}(\mathrm{wt}(\gamma_0)) + \delta_0(J) = w(I, i_0) + w.$$

Now consider paths $\eta \in W_n(I)$ such that $\eta = \gamma_0 \eta_1$ for some path η_1 with $s(\eta_1) = J$. The sum of the masses of all such paths equals

$$\mathrm{wt}(\gamma_0) r(J) p^{n-l}$$

and

$$\mathrm{Log}(\mathrm{wt}(\gamma_0) r(J) p^{n-l}) \subset \mathrm{Log}(\mathrm{wt}(\gamma_0) y^{\delta_0(J)} p^L p^{n-l})$$
$$= \mathrm{Log}(y^{w(I,i_0)} y^w p^{n+L-l})$$
$$\subset \mathrm{Log}(p^{n+L}) + w(I, i_0).$$

Using (6.1) we deduce that, for large n, every near-F_0 path $\eta \in W_n(I)$ of the form $\eta = \gamma_0 \eta_1$ has

$$\mathrm{Log}(\mathrm{mass}(\eta)) \subset \mathrm{Log}(p^n) + w(I, i_0).$$

Provided $\eta \notin \{\gamma^{(I,i)} : 1 \leq i \leq \rho(I)\}$, we assign any such near-F_0 path η to the set (I, i_0). In particular, we assign γ to (I, i_0).

We can similarly assign all near-F_∞ paths to the sets (I,i), $1 \leq i \leq \rho(I)$, and make sure that each (I,i) contains a near-F_∞ path $\eta^{(I,i)}$ with $\delta_\infty(\mathrm{mass}(\eta^{(I,i)})) = w(I,i) + n\delta$ and that $\mathrm{Log}(\mathrm{mass}(\eta)) \subset \mathrm{Log}(y^{w(I,i)} p^n)$ whenever η is a near-F_∞ path in (I,i). (Recall that, by our choice of n, the sets of near-F_0 paths and near-F_∞ paths are disjoint.) Once the near-F_0 and near-F_∞ paths have been allocated, it is easy to assign the remaining paths γ with $s(\gamma) = I$, $l(\gamma) = n$ to the sets (I,i) in such a way that $\mathrm{Log}(\mathrm{mass}(\gamma)) \subset \mathrm{Log}(y^{w(I,i)} p^n)$ for all $\gamma \in (I,i)$. This leads us to:

PROPOSITION 6.7. *For n large enough, A^n may be state-split in such a way that the resulting matrix $\widehat{A^n}$ has a right eigenvector $\widehat{r}$ which totally conforms to p^n.*

PROOF. It is clear that the right eigenvector $\widehat{r}$, with

$$\widehat{r}(I,i) = \sum_{\gamma \in (I,i)} \mathrm{mass}(\gamma),$$

resulting from the above state-splitting has all the desired properties, except perhaps (c). We complete the proof by verifying (c). Put $B = \widehat{A^n}$ and let $\psi : (X_B, \sigma_B) \to (X_{A^n}, \sigma_{A^n})$ be the canonical block isomorphism; ψ is a 1-block map arising from a graph homomorphism whose underlying state map sends each (I,i) to I. (See the beginning of section 5.) Let F be a face of $W(p)$ and $\widehat{\mathcal{C}}$ a principal F-component of B. By (3.13) of [**MT1**], ψ sends $\widehat{\mathcal{C}}$ onto a principal F-component of A^n, which is an irreducible component of the n-th power of a principal F-component $\mathcal{C}$ of A. As in (3.1), we assume without loss of generality that $p_F \in R(\Delta(p_F))$, the matrix $A_{\mathcal{C}}$ representing $\mathcal{C}$ is over $R(\Delta(p_F))$, and that the vector $r_{\mathcal{C}}$ (with $r_{\mathcal{C}}(I) = r(I)_F$ for $I \in \mathcal{C}$) is over $R(\Delta(p_F))$. Then the matrix $B_{\widehat{\mathcal{C}}}$ representing $\widehat{\mathcal{C}}$ is over $R(\Delta(p_F))$ and we have $\Delta(B_{\widehat{\mathcal{C}}}) = \Delta(A_{\mathcal{C}})$, $c_{B_{\widehat{\mathcal{C}}}} \Delta(B_{\widehat{\mathcal{C}}}) = c_{A_{\mathcal{C}}} \Delta(A_{\mathcal{C}})$. Let d be the period of $B_{\widehat{\mathcal{C}}}$ and write $c = c_{B_{\widehat{\mathcal{C}}}}$. Let D be a diagonal matrix with monomials for its diagonal entries such that the matrix

$$\frac{1}{c^{1/d}} D (A_{\mathcal{C}})^n D^{-1}$$

is over $R(\Delta(A_{\mathcal{C}}))$. Put $\widetilde{r}_{\mathcal{C}} = D r_{\mathcal{C}}$. The map ψ lifts D to a diagonal matrix $\widetilde{D}$ such that $\widetilde{D}((I,i),(I,i)) = D(I,I)$ for $(I,i) \in \mathcal{C}$ and

$$\frac{1}{c^{1/d}} \widetilde{D} B_{\widehat{\mathcal{C}}} \widetilde{D}^{-1}$$

is over $R(\Delta(A_{\mathcal{C}})) = R(\Delta(B_{\widehat{\mathcal{C}}}))$. Let $\widehat{C}$ be a period class of $\widehat{\mathcal{C}}$. If (I, i_0) is in $\widehat{C}$ for some i_0, we consider

$$T(I) = \{1 \le i \le \rho(I) : (I,i) \text{ contains a path in } G(A_{\mathcal{C}})\}.$$

It is easy to see that then (I,i) is in $\widehat{C}$ for every $i \in T(I)$. It follows from (5) of [**T3**] that

$$\widehat{r}(I,i)_F = \sum_{\gamma} \mathrm{wt}_{A_{\mathcal{C}}}(\gamma)\, r(t(\gamma))_F,$$

where the sum is over all $G(A_{\mathcal{C}})$-paths $\gamma \in (I,i)$, and that

$$p_F^n\, r_{\mathcal{C}}(I) = \sum_{i \in T(I)} \widehat{r}(I,i)_F.$$

For $(I, i_0) \in \widehat{C}$ we then have

$$p_F^n\, \widetilde{r}_{\mathcal{C}}(I) = \sum_{i \in T(I)} D(I,I)\, \widehat{r}(I,i)_F.$$

Since p_F^n is invertible in R_{p_F} and $\{I : (I,i) \in \widehat{C} \text{ for some } i\}$ is a union of period classes of $\mathcal{C}$, it follows that the $R(\Delta(B_{\widehat{C}}))\left[\frac{1}{(p_F)^d}\right]$-module

$$\langle D(I,I)\widehat{r}(I,i)_F : (I,i) \in \widehat{C} \rangle$$

equals R_{p_F}. □

Splitting the conforming eigenvector in the one-variable case

We continue with the proof of (3.2) in the case of a single variable. As in section 6, $R = \mathbb{Z}[y^{\pm}]$, the matrix A is over R^+ and irreducible, $\beta_A = p \in R^+$, and r is a right eigenvector of A satisfying (a)–(c). For $q \in R$ we have $\delta_0(q) = \min(\mathrm{Log}(q))$ and $\delta_\infty(q) = \max(\mathrm{Log}(q))$. We simplify our notation by writing $\delta_0 = \delta_0(p)$, $\delta_\infty = \delta_\infty(p)$ and, for $I \in S$, letting $\delta_0(I) = \delta_0(r(I))$, $\delta_\infty(I) = \delta_\infty(r(I))$. The face F_0 of $W(p)$ is exposed upon multiplication by 1; we write A_{F_0} and r_{F_0} for the corresponding faces of A and r. Similarly, F_∞ is exposed upon multiplication by -1, and we denote the corresponding faces of A and r by A_{F_∞} and r_{F_∞}. For the face $F = W(p)$ of $W(p)$, we put $A_F = A$, $r_F = r$.

Using [**PS, MT1**], we divide A, p by c_A and conjugate A by a suitable diagonal matrix to assume A is over $R(\Delta(A))^+$ and $p \in R(\Delta(p))^+$. We also assume that $\mathrm{Log}(p)$ generates $\mathbb{Z}$; that is, $R(\Delta(p)) = R$. Replacing A by a suitable power, (6.7) and (5.3) enable us to assume, without loss of generality, that r totally conforms to p and that every F-component of A is aperiodic for every $F \in \mathcal{F}(p)$.

We will use the following three estimates involving the coefficients of p^n. They are special cases of results of [**MT4**] which will be stated in full generality in section 9. Proposition (7.1) is the same as (6.3).

PROPOSITION 7.1. *Let $L \in \mathbb{N}$ and $K > 0$. There exist $\epsilon > 0$ and $N \in \mathbb{N}$ such that for $a \in \mathbb{Z}$ and nonzero $b \in \mathrm{Log}(p^L) - L\delta_0$ the inequality*

$$(p^n)_{a+b} \geq K(p^n)_a$$

holds whenever $n \geq N$ and $n\delta_0 \leq a \leq n\delta_0 + \epsilon n$.

PROPOSITION 7.2. *Let $\epsilon, \theta > 0$ and $D \in \mathbb{N}$. There exists $N \in \mathbb{N}$ such that*

$$\left| \frac{(p^n)_{a+b-w}}{(p^n)_{a+b-w'}} \cdot \frac{(p^n)_{a+b'-w'}}{(p^n)_{a+b'-w}} - 1 \right| < \theta$$

whenever $n \geq N$, $n\delta_0 + \epsilon n \leq a \leq n\delta_\infty - \epsilon n$ and $|b|, |b'|, |w|, |w'| \leq D$.

PROPOSITION 7.3. *Let $\epsilon > 0$, $l \in \mathbb{Z}$ and $D \in \mathbb{N}$. There exist $K, N \in \mathbb{N}$ such that*

$$\frac{1}{K} \leq (p^n)_{a+b}/(p^{n+l})_a \leq K$$

whenever $n \geq N$, $n\delta_0 + \epsilon n \leq a \leq n\delta_\infty - \epsilon n$ and $|b| \leq D$.

7. SPLITTING THE EIGENVECTOR IN THE ONE-VARIABLE CASE

Fix $I \in S = S(G(A))$. As before, write $r(I) = \sum_{i=1}^{\rho(I)} y^{w(I,i)}$. For $n \in \mathbb{N}$ and $a \in \text{Log}(p^n r(I))$, let $\pi(n, a)$ be the probability vector of length $\rho(I)$ whose i-th entry is given by

$$\pi(n, a, i) = (y^{w(I,i)} p^n)_a / (p^n r(I))_a.$$

LEMMA 7.4. *Let $\epsilon, \theta > 0$ and $D, l \in \mathbb{Z}^+$. There exists $N \in \mathbb{N}$ such that*

$$\left| \frac{\pi(n+l, a+b, i)}{\pi(n, a, i)} - 1 \right| < \theta$$

whenever $n \geq N$, $n\delta_0 + \epsilon n \leq a \leq n\delta_\infty - \epsilon n$ and $|b| \leq D$.

PROOF. We have

$$(y^{w(I,i)} p^{n+l})_{a+b} = \sum_{b' \in \text{Log}(p^l)} (y^{w(I,i)} p^n)_{a+b-b'} (p^l)_{b'}$$

$$= \sum_{b' \in \text{Log}(p^l)} \pi(n, a+b-b', i)(p^n r(I))_{a+b-b'} (p^l)_{b'}.$$

Since (7.2) tells us that the ratio of each $\pi(n, a+b-b', i)$ to $\pi(n, a, i)$ may be made arbitrarily close to 1 by increasing n, the above sum lies between

$$(1-\theta)\pi(n, a, i) \sum_{b' \in \text{Log}(p^l)} (p^n r(I))_{a+b-b'} (p^l)_{b'} = (1-\theta)\pi(n, a, i)(p^{n+l} r(I))_{a+b}$$

and

$$(1+\theta)\pi(n, a, i)(p^{n+l} r(I))_{a+b}$$

for all large n. □

Let $k_1 = |\Delta(p)/\Delta(A)|$. Using condition (c) and the fact that r totally conforms to p, we find $L \in \mathbb{Z}^+$, $q^{F_0}(J) \in \mathbb{Z}$, $q^{F_\infty}(J) \in \mathbb{Z}$ and, for $0 \leq u < k_1$, polynomials $q^{(u)}(J) \in R(\Delta(A))$ such that:

(1) $\text{Log}(r(J)) - \delta_0(J) + L\delta_0 \subset \text{Log}(p^L)$ and $\text{Log}(r(J)) - \delta_\infty(J) + L\delta_\infty \subset \text{Log}(p^L)$ for every $J \in S$,
(2) $\sum_{J \in S} q^{(u)}(J) r(J) = y^u p^L$,
(3) $\sum_{J \in \mathcal{C}} q^{F_0}(J) r(J)_{\delta_0(J)} = (p_{\delta_0})^L$ for each principal F_0-component $\mathcal{C}$ of A,
(4) $\sum_{J \in \mathcal{C}} q^{F_\infty}(J) r(J)_{\delta_\infty(J)} = (p_{\delta_\infty})^L$ for each principal F_∞-component $\mathcal{C}$ of A,
(5) if $F \in \mathcal{F}(p)$ and $\mathcal{C}$ is a principal F-component then, for any $J, J' \in \mathcal{C}$, there exists an A_F-path of length $L - |S|$ from J to J'.

We fix an even number D such that $D \geq 2k_1$ and

$$D > 2\max\{|w| : w \in \text{Log}(p^L) \cup \bigcup_{\substack{J \in S \\ 0 \leq u < k_1}} \text{Log}(q^{(u)}(J)) \cup \bigcup_{J \in S} \text{Log}(r(J))\}.$$

We also pick a number M which dominates absolute values of all the constants and coefficients appearing in (2)–(4); that is, M equals at least the maximum of the set

$$\{(p^L)_w, r(J)_w, |q^{(u)}(J)_w|, |q^{F_0}(J)|, |q^{F_\infty}(J)| : J \in S, 0 \leq u < k_1, |w| < D/2\}.$$

For $J \in S$ and $w \in \Delta(A)$, let
$$W_n(I, J) = \{\gamma \in W_n(I) : t(\gamma) = J\},$$
$$W_n(I, J, w) = \{\gamma \in W_n(I, J) : \mathrm{Log}(\mathrm{wt}(\gamma)) = w\}.$$
We say that $\gamma \in W_n(I, J, w)$ is of *type* (n, J, w). For $W \subset W_n(I)$, we write
$$\mathrm{mass}(W) = \sum_{\gamma \in W} \mathrm{mass}(\gamma).$$

Recall that our goal is to partition $W_n(I)$ into sets (I, i), $1 \le i \le \rho(I)$, so that $\mathrm{mass}((I,i)) = y^{w(I,i)} p^n$. We assume $\rho(I) \ge 2$, for otherwise we would have nothing to do. Equations (2)–(4) will enable us to correct distributions that come close to our goal. The next two propositions will serve to provide us with the paths we will use in making corrections.

PROPOSITION 7.5. *There exist $\epsilon > 0$ and $N \in \mathbb{N}$ so that for each $n \ge N$ and $a \in [\delta_0(I) + n\delta_0, \delta_0(I) + n\delta_0 + \epsilon n]$ we have*
  (i) $\mathrm{mass}\left(\{\gamma \in W_n(I) : \delta_0(\mathrm{mass}(\gamma)) = a\}\right)_a \ge \frac{1}{2}(p^n r(I))_a$,
  (ii) *a state $J_0 \in S$ with*
$$|W_n(I, J_0, a - \delta_0(J_0))| \ge \frac{1}{2M^2 |S|} (p^n r(I))_a,$$
  (iii) *a principal F_0-component $\mathcal{C}$ of A and, for every $J \in \mathcal{C}$, a path $\eta^J \in W_L(J_0, J)$ such that*
$$\{\gamma \eta^J : \gamma \in W_n(I, J_0, a - \delta_0(J_0))\} \subset W_{n+L}(I, J, a + L\delta_0 - \delta_0(J)).$$

PROOF. Letting $M'_n(a) = \{\gamma \in W_n(I) : \delta_0(\mathrm{mass}(\gamma)) = a\}$ and
$$M''_n(a) = \{\gamma \in W_n(I) : a - D \le \delta_0(\mathrm{mass}(\gamma)) \le a - 1 \text{ and } \mathrm{mass}(\gamma)_a \ne 0\},$$
we have
$$(p^n r(I))_a = \mathrm{mass}(M'_n(a))_a + \mathrm{mass}(M''_n(a))_a.$$
As a result of (1), the mass of each element of $M''_n(a)$ contributes to $(p^n r(I))_{a-j}$ for some nonzero $j \in \mathrm{Log}(p^L) - L\delta_0$. Making at most D applications of (7.1) with $K = 2MD$, we find $\epsilon > 0$ such that
$$\mathrm{mass}(M''_n(a))_a \le \frac{1}{2}(p^n r(I))_a$$
for all large n and $a \in [\delta_0(I) + n\delta_0, \delta_0(I) + n\delta_0 + \epsilon n]$. This implies (i) and that
$$|M'_n(a)| \ge \frac{1}{2M}(p^n r(I))_a.$$
Hence, we can find $J_0 \in S$ with
$$|W_n(I, J_0, a - \delta_0(J_0))| = |\{\gamma \in W_n(I, J_0) : \delta_0(\mathrm{mass}(\gamma)) = a\}| \ge \frac{1}{2M|S|}(p^n r(I))_a.$$
Since $(2M|S|)^{-1} \ge (2M^2|S|)^{-1}$, we have (ii). Considering A_{F_0}-paths of length $|S|$ starting at J_0, we find a principal F_0-component $\mathcal{C}$ so that there exists an A_{F_0}-path η of length $|S|$ from J_0 to a state of $\mathcal{C}$. Following η by A_{F_0}-paths of length $L - |S|$

and using (5), we obtain A_{F_0}-paths $\eta^J \in W_L(J_0, J)$. Since η^J is an A_{F_0}-path, we have
$$\delta_0(J_0) + L\delta_0 = \text{Log}(\text{wt}(\eta^J)) + \delta_0(J),$$
which implies (iii). □

PROPOSITION 7.6. *For every $\epsilon > 0$ and $\bar{D} \in \mathbb{N}$, there exist $N, \bar{K} \in \mathbb{N}$ such that*
$$|W_n(I, J, a+b)| = (A^n(I, J))_{a+b} \geq \frac{1}{\bar{K}}(p^n r(I))_a$$
whenever $n \geq N$, $a \in [n\delta_0 + \epsilon n, n\delta_\infty - \epsilon n]$, $y^{a+b} \in \Delta(A)$, $|b| \leq \bar{D}$, and $I, J \in S$.

LEMMA 7.7. *There exist $J_1 \in S$, $y^{w_1} \in \Delta(A)$ and $l_1 \in \mathbb{N}$ such that, for every $h \in \mathbb{N}$, a translate of $\{b \in \mathbb{Z} : y^b \in \Delta(A), |b| \leq h\}$ is realized in the following way: For every $y^b \in \Delta(A)$ with $|b| \leq h$, there exists $\gamma_b \in W_{l_1 h}(J_1, J_1)$ with $\text{wt}(\gamma_b) = y^{hw_1 + b}$.*

PROOF. Since $k_1 = |\Delta(p)/\Delta(A)|$, the group $\Delta(A)$ is generated by y^{k_1}. Find cycles γ, η such that $l(\gamma) = l(\eta)$, $\text{wt}(\gamma)/\text{wt}(\eta) = y^{k_1}$ and γ, η pass through a common state J_1. Let $l_1 = l(\eta\gamma)$ and $w_1 = \text{Log}(\text{wt}(\eta\gamma))$. For $h \in \mathbb{N}$ the desired cycles are then among
$$\eta^{2h}, \eta^{2h-1}\gamma, \ldots, \eta^h \gamma^h, \ldots, \gamma^{2h}.$$
□

PROOF OF (7.6). By (5), A^L has no zero entries. For $J, J' \in S$, fix $\eta^{JJ'} \in W_L(J, J')$. We shall apply (7.7) with
$$h = D + 2\max\{|\text{Log}(\text{wt}(\eta^{JJ'}))| : J, J' \in S\}.$$
Let $l = 2L + l_1 h$. The equation $A^{n-l} r = p^{n-l} r$ implies the existence of $J_0 \in S$ and $b' \in \text{Log}(r(J_0))$ such that $y^{b'} \in \Delta(A)$ and
$$(p^{n-l} r(I))_{a+b-hw_1} \leq MD|S| (A^{n-l}(I, J_0))_{a+b-hw_1-b'}.$$
Using (7.7) to find $\eta \in W_{l_1 h}(J_1, J_1)$ with
$$\text{wt}(\eta) = y^{hw_1 + b'} / \text{wt}(\eta^{J_0 J_1} \eta^{J_1 J}),$$
and following each element of $W_{n-l}(I, J_0, a+b-hw_1-b')$ with the path $\eta^{J_0 J_1} \eta \eta^{J_1 J}$, we deduce that
$$(p^{n-l} r(I))_{a+b-hw_1} \leq MD|S|(A^n(I, J))_{a+b}.$$
Finally we apply (7.3) to see that $(p^n r(I))_a$ is less than a constant multiple of $(p^{n-l} r(I))_{a+b-hw_1}$ for all large n and $n\delta_0 + \epsilon n \leq a \leq n\delta_\infty - \epsilon$. □

To describe the application of (7.6) and the equation (2) to the correction of the midrange, we consider $\epsilon, \xi > 0$, integers a_0^*, a_∞^* with $n\delta_0 + \epsilon n \leq a_0^* < a_\infty^* \leq n\delta_\infty - \epsilon n$ and a partition of $W_n(I)$ into $\rho(I)$ sets $P(n, i)$ satisfying the following two conditions.

(i) For $\delta_0(I) + n\delta_0 \leq a \leq a_0^*$ and for $a_\infty^* \leq a \leq \delta_\infty(I) + n\delta_\infty$ we have
$$\mathrm{mass}(P(n,i))_a = (y^{w(I,i)}p^n)_a.$$

(ii) For every type (n, J, w) with $a_0^* - 3D < w < a_\infty^* + 3D$ we have
$$1 - \xi \leq \frac{|W_n(I, J, w) \cap P(n, i)|}{|W_n(I, J, w)|} \cdot \frac{1}{\pi(n, w + \delta_0(J), i)} \leq 1 + \xi.$$

LEMMA 7.8. *For every $\epsilon \in (0, \frac{1}{2})$, there exist $\xi > 0$ and $N \in \mathbb{N}$ such that, for any $n \geq N$ and integers a_0^*, a_∞^* with $n\delta_0 + \epsilon n \leq a_0^* < a_\infty^* \leq n\delta_\infty - \epsilon n$, the existence of a partition of $W_n(I)$ into $\rho(I)$ sets $P(n, i)$ satisfying (i) and (ii) implies the existence of a partition of $W_{n+L}(I)$ into $\rho(I)$ sets $P(n + L, i)$ such that $\mathrm{mass}(P(n + L, i)) = y^{w(I,i)}p^{n+L}$.*

PROOF. Let $\widetilde{P}(n+L, i)$ be the set of paths of length $n+L$ obtained by extending the paths in $P(n, i)$:
$$\widetilde{P}(n + L, i) = \{\gamma\eta : \gamma \in P(n, i), \eta \in W_L(t(\gamma))\}.$$

Clearly $\mathrm{mass}(\widetilde{P}(n + L, i)) = \mathrm{mass}(P(n, i))p^L$. Consider $a \in (a_0^*, a_\infty^*)$. Picking $0 \leq u < k_1$ with $y^u \Delta(A) = y^a \Delta(A)$ and multiplying the equation (2) by y^{a-u},
$$y^a p^L = \sum_{J \in S} y^{a-u} q^{(u)}(J) r(J)$$
$$= \sum_{J \in S} \sum_{|b| < D/2} q^{(u)}(J)_b \, y^{a-u+b} r(J).$$

This enables us to "correct" the $\widetilde{P}(n+L, i)$ by exchanging paths: Consider moving, for each $J \in S$ and $|b| < D/2$, exactly $q^{(u)}(J)_b$ elements of $W_{n+L}(I, J, a - u + b)$ from $\widetilde{P}(n + L, j)$ to $\widetilde{P}(n + L, i)$. (If $q^{(u)}(J)_b$ is negative, we move the paths in the opposite direction, that is, from $\widetilde{P}(n + L, i)$ to $\widetilde{P}(n + L, j)$.) According to the above equation, the net result of this operation is the addition of $y^a p^L$ to $\mathrm{mass}(\widetilde{P}(n + L, i)) = \mathrm{mass}(P(n, i))p^L$ and the subtraction of $y^a p^L$ from $\widetilde{P}(n + L, j)$. So, if each $\widetilde{P}(n + L, i)$ contains the paths required to repeat this procedure $(y^{w(I,i)}p^n)_a - \mathrm{mass}(P(n, i))_a$ times, we will have proved the lemma. (Note that, since $\sum_j (y^{w(I,j)}p^n - \mathrm{mass}(P(n, j))) = 0$, any time we need to add/subtract $y^a p^L$ for i we can find j for which we need to subtract/add $y^a p^L$.) How many elements of each type do we need in $\widetilde{P}(n+L, i)$ in order to be sure of making all the corrections? We will estimate this number in terms of ξ and show that ξ can be chosen to ensure the availability of the desired paths. First note that

$$\mathrm{mass}(P(n, i))_a = \sum_J \sum_{b \in \mathrm{Log}(r(J))} |W_n(I, J, a - b) \cap P(n, i)| \, r(J)_b$$
$$\leq (1 + \xi) \sum_J \sum_{b \in \mathrm{Log}(r(J))} \pi(n, a - b + \delta_0(J), i) \, |W_n(I, J, a - b)| \, r(J)_b$$
$$\leq (1 + 2\xi) \sum_J \sum_{b \in \mathrm{Log}(r(J))} \pi(n, a, i) \, |W_n(I, J, a - b)| \, r(J)_b,$$

7. SPLITTING THE EIGENVECTOR IN THE ONE-VARIABLE CASE

where the second inequality results from (7.4) by making sure n is large enough. Since the last double sum equals $(y^{w(I,i)}p^n)_a$, we find that

$$\mathrm{mass}(P(n,i))_a \leq (1+2\xi)(y^{w(I,i)}p^n)_a.$$

A similar argument shows that $\mathrm{mass}(P(n,i))_a \geq (1-2\xi)(y^{w(I,i)}p^n)_a$. Thus we have

$$(*) \qquad -2\xi(r(I)p^n)_a \leq \mathrm{mass}(P(n,i))_a - (y^{w(I,i)}p^n)_a \leq 2\xi(r(I)p^n)_a.$$

for all large n. Observe that each type $(n+L, J, w)$ is called upon in correcting at most $2D$ coefficients of $\mathrm{mass}(P(n,i))$ and at each step of the correction we need at most M paths of type $(n+L, J, w)$. Combining these observations with $(*)$ we see that the existence, for $w \in (a_0^* - D, a_\infty^* + D)$, of

$$2DM(2\xi)\max\{(p^n r(I))_{w+b} : |b| < D\}$$

elements of $W_{n+L}(I, J, w)$ in each $\widetilde{P}(n+L, i)$ is sufficient to make all the corrections. Use (7.3) to find $\mu > 0$ such that $\mu < \pi(n, a, i)$ for all $a \in [a_0^* - 3D, a_\infty^* + 3D]$ and all large n. Let $\bar{K}$ be as in (7.6) for $\bar{D} = 5D$. Take

$$\xi = \mu/(8DM\bar{K}).$$

For $J \in S$, fix $J_0 \in S$ and a path $\eta \in W_L(J_0, J)$, and put $w_0 = \mathrm{Log}(\mathrm{wt}(\eta))$. Since $\mathrm{Log}(w_0 r(J)) \subset \mathrm{Log}(r(J_0)p^L)$, we have $|w_0| < 3D/2$. It follows from this and (7.6) that

$$\frac{1}{\bar{K}}\max\{(p^n r(I))_{w+b} : |b| < D\} \leq |W_n(I, J_0, w - w_0)|.$$

Since $\xi < \frac{1}{2}$,

$$\frac{1}{\bar{K}}\mu(1-\xi) \geq \frac{\mu}{2\bar{K}} = 4DM\xi.$$

Hence, for $w \in (a_0^* - D, a_\infty^* + D)$ we have

$$|W_{n+L}(I, J, w) \cap \widetilde{P}(n+L, i)|$$

$$\geq |W_n(I, J_0, w - w_0)) \cap P(n, i)|$$

$$\geq (1-\xi)|W_n(I, J_0, w - w_0))|\pi(n, w - w_0 + \delta_0(J), i)$$

$$\geq \frac{(1-\xi)\mu}{\bar{K}}\max\{(p^n r(I))_{w+b} : |b| < D\}$$

$$\geq 4DM\xi \max\{(p^n r(I))_{w+b} : |b| < D\}. \qquad \square$$

Clearly there are analogues of (7.1) and (7.5) for the high end of $\mathrm{Log}(p^n)$ and $\mathrm{Log}(p^n r(I))$. (These are established similarly, or just by making the change of variable $y \mapsto y^{-1}$.) Fix

$$K > (4M^4 D^4)^D.$$

Find $\epsilon_0 \in (0, \frac{1}{4})$ so that (7.1), (7.5) and their high end counterparts hold with $\epsilon = \epsilon_0$. In particular, for all large n we have $(p^n)_{a+b} \geq K(p^n)_a$ when $a \in [\delta_0(I) + n\delta_0, \delta_0(I) + n\delta_0 + \epsilon_0 n]$ and nonzero $b \in \mathrm{Log}(p^L) - L\delta_0$, as well as when $a \in [\delta_\infty(I) + n\delta_\infty - \epsilon_0 n, \delta_\infty(I) + n\delta_\infty]$ and nonzero $b \in \mathrm{Log}(p^L) - L\delta_\infty$. We will make use of

(7.4), (7.6) and (7.8) with $\epsilon = \epsilon_0/(L+4)$. Let $\xi > 0$ be as determined by (7.8) for $\epsilon = \epsilon_0/(L+4)$. Let $\epsilon_1 = 2\epsilon_0/(L+4)$. For large n, we will construct a partition of $W_n(I)$ to which (7.8) can be applied. We describe the construction in detail at the low end.

We will use induction on a, starting with $a = 0$ and going up to a value slightly larger than $\epsilon_1 n$. At each step of the induction we will increase n by L in order to be able to use (3) to make corrections. Put

$$\widetilde{a} = \delta_0(I) + (n + aL)\delta_0 + a\,.$$

At the start of the a-th step we will have

$$\{\gamma \in W_{n+aL}(I) : \delta_0(\mathrm{mass}(\gamma)) < \widetilde{a}\}$$

partitioned into $\rho(I)$ sets $P(n, a, i)$ with $\mathrm{mass}(P(n, a, i))_b = (y^{w(I,i)} p^{n+aL})_b$ for all $b < \widetilde{a}$. We will then allocate those $\gamma \in W_{n+aL}(I)$ with $\delta_0(\mathrm{mass}(\gamma)) = \widetilde{a}$, extend by L, use (3) to correct and end up having

$$\{\gamma \in W_{n+(a+1)L}(I) : \delta_0(\mathrm{mass}(\gamma)) < \widetilde{a+1}\}$$

partitioned into sets $P(n, a+1, i)$ with $\mathrm{mass}(P(n, a+1, i))_b = (y^{w(I,i)} p^{n+(a+1)L})_b$ for all $b < \widetilde{a+1}$.

Recall from section 6 that $(p^n r(I))_{\delta_0(I)+n\delta_0+a} \to \infty$ as $n \to \infty$ except where $p_{\delta_0} = 1$ and a is such that

$$(*) \qquad (p^l r(I))_{\delta_0(I)+l\delta_0+a} = (p_{\delta_0})^l\, r(I)_{\delta_0(I)+a} = r(I)_{\delta_0(I)+a} \text{ for all } l.$$

Note that $(*)$ holds for at most finitely many a; in fact, $(*)$ implies that $a \leq H$, where H is as in (6.1). Such a are easy to deal with: Recall from (6.6) that in this case there are exactly $r(I)_{\delta_0(I)+a}$ paths $\gamma \in W_{n+aL}(I)$ with $\delta_0(\mathrm{mass}(\gamma)) = \delta_0(I) + (n+aL)\delta_0 + a$, and observe that for such γ no element of $\mathrm{Log}(\mathrm{mass}(\gamma)) - \delta_0(I) - (n+aL)\delta_0$ other than a satisfies $(*)$. We will bijectively assign these $r(I)_{\delta_0(I)+a}$ paths to the $r(I)_{\delta_0(I)+a}$ sets corresponding to the values of i with $w(I, i) = \delta_0(I) + a$. Outside of this exceptional case, we will assign paths by type. For each $W_{n+aL}(I, J, w)$ with $w + \delta_0(J) = \widetilde{a} = \delta_0(I) + (n+aL)\delta_0 + a$ we will employ the probability vector $\pi(n + aL, \widetilde{a})$ and want to place

$$(**) \qquad \pi(n + aL, \widetilde{a}, i)\, |W_{n+aL}(I, J, w)|$$

elements of $W_{n+aL}(I, J, w)$ in the i-th set. We will indicate this by saying that $W_{n+aL}(I, J, w)$ is allocated according to $\pi(n + aL, \widetilde{a})$: Since the number $(**)$ need not be integral, what we mean by this is that the number placed in the i-th set differs from $(**)$ by less than 1.

In order to proceed by induction, suppose a is such that all paths $\gamma \in W_{n+aL}(I)$ with $\delta_0(\mathrm{mass}(\gamma)) < \widetilde{a} = \delta_0(I) + (n+aL)\delta_0 + a$ have been allocated to form $\rho(I)$ disjoint sets $P(n, a, i)$ satisfying

$$\mathrm{mass}(P(n, a, i))_b = (y^{w(I,i)} p^{n+aL})_b$$

for all $b < \widetilde{a}$. (When $a = 0$, we start with having allocated nothing.) Note that the set $\{\gamma \in W_{n+aL}(I) : \delta_0(\mathrm{mass}(\gamma)) < \widetilde{a}\}$ is the union of $W_{n+aL}(I, J, w)$ over (J, w) with $w < \widetilde{a} - \delta_0(J)$. If a satisfies $(*)$, assign the $r(I)_{\delta_0(I)+a}$ paths

$\gamma \in W_{n+aL}(I)$ with $\delta_0(\text{mass}(\gamma)) = \widetilde{a}$ as indicated above. Otherwise, allocate each $W_{n+aL}(I, J, w)$ with $w = \widetilde{a} - \delta_0(J)$ according to $\pi(n+aL, \widetilde{a})$. Denote by $P'(n,a,i)$ the set containing these allocations and $P(n,a,i)$. Use (6.1) to observe that (1) guarantees that $\text{Log}(\text{mass}(\gamma)) \subset \text{Log}(y^{w(I,i)}p^{n+aL})$ whenever $\gamma \in P'(n,a,i)$. For each $1 \le i \le \rho(I)$, let

$$\widetilde{P}(n, a+1, i) = \{\gamma\eta : \gamma \in P'(n,a,i), \eta \in W_L(t(\gamma)), \delta_0(\text{mass}(\gamma\eta)) \le \widetilde{a} + L\delta_0\}.$$

It is easy to see that $\text{mass}(P'(n,a,i))_b = (y^{w(I,i)}p^{n+aL})_b$ for $b < \widetilde{a}$. Hence,

$$\text{mass}(\widetilde{P}(n, a+1, i))_b = (y^{w(I,i)}p^{n+aL+L})_b$$

for $b < \widetilde{a} + L\delta_0 = \widetilde{a+1} - 1$. In the case $b = \widetilde{a} + L\delta_0 = \widetilde{a+1} - 1$,

$$(***) \quad \begin{aligned} \text{mass}(\widetilde{P}(n, a+1, i))_{\widetilde{a}+L\delta_0} &- (y^{w(I,i)}p^{n+aL+L})_{\widetilde{a}+L\delta_0} \\ &= \left(\text{mass}(P'(n,a,i))_{\widetilde{a}} - (y^{w(I,i)}p^{n+aL})_{\widetilde{a}}\right)(p_{\delta_0})^L. \end{aligned}$$

If a satisfies $(*)$, we have no corrections to make: We simply take $P(n,a+1,i) = \widetilde{P}(n,a+1,i)$ and find that

$$(\dagger) \quad \text{mass}(P(n,a+1,i))_b = (y^{w(I,i)}p^{n+(a+1)L})_b$$

for all $b < \widetilde{a+1}$. Suppose a does not satisfy $(*)$. Then

$$(p^{n+aL}r(I))_{\widetilde{a}} \to \infty$$

as $n \to \infty$. Provided $a \le \epsilon_0 n$, we can apply (7.5) and take $J_0 \in S$ and $\mathcal{C}$ to be as provided by (7.5) for $n+aL$ and $\widetilde{a}$: We have

$$(\dagger\dagger) \quad |W_{n+aL}(I, J_0, \widetilde{a} - \delta_0(J_0))| \ge \frac{1}{2M^2|S|}(p^{n+aL}r(I))_{\widetilde{a}},$$

and for every $J \in \mathcal{C}$ we have a path $\eta^J \in W_L(J_0, J)$ such that $\delta_0(\text{mass}(\gamma\eta^J)) = \widetilde{a} + L\delta_0$ whenever $\gamma \in W_{n+aL}(I, J_0, \widetilde{a} - \delta_0(J_0))$. With this choice of $\mathcal{C}$, multiply (3) by $y^{\widetilde{a}+L\delta_0}$ to see that

$$(p_{\delta_0})^L = \left(\sum_{J \in \mathcal{C}} q^{F_0}(J) y^{\widetilde{a}+L\delta_0 - \delta_0(J)} r(J)\right)_{\widetilde{a}+L\delta_0}.$$

Consider moving, for each $J \in \mathcal{C}$, exactly $q^{F_0}(J)$ elements of $W_{n+(a+1)L}(I, J, \widetilde{a} + L\delta_0 - \delta_0(J))$ from $\widetilde{P}(n, a+1, j)$ to $\widetilde{P}(n, a+1, i)$. (If $q^{F_0}(J) < 0$, we move the paths in the opposite direction.) The above equation means that this will increase $\text{mass}(\widetilde{P}(n,a+1,i))_{\widetilde{a}+L\delta_0}$ by $(p_{\delta_0})^L$ and decrease $\text{mass}(\widetilde{P}(n,a+1,j))_{\widetilde{a}+L\delta_0}$ by $(p_{\delta_0})^L$. Considering $(***)$, we will complete the inductive step by showing that each $\widetilde{P}(n, a+1, i)$ contains the paths needed to apply this procedure

$$(\dagger\dagger\dagger) \quad (y^{w(I,i)}p^{n+aL})_{\widetilde{a}} - \text{mass}(P'(n,a,i))_{\widetilde{a}}$$

times. The following lemma will be used in estimating $(\dagger\dagger\dagger)$.

LEMMA 7.9. *For $a \le \frac{3}{2}\epsilon_1 n$ we have*

$$\sum_{\substack{u \in \text{Log}(p^L) - L\delta_0, \\ u \ne 0}} (y^{w(I,i)}p^{n+aL})_{\widetilde{a}-u} \le \frac{D}{K}\pi(n+aL, \widetilde{a}, i)(r(I)p^{n+aL})_{\widetilde{a}}.$$

PROOF. Since
$$\frac{\frac{3}{2}\epsilon_1 n}{n + \frac{3}{2}\epsilon_1 nL} \leq \frac{3}{2}\epsilon_1 = \frac{3\epsilon_0}{L+4} \leq \frac{3}{4}\epsilon_0,$$

(7.1) applies to the terms of the above sum and shows that the sum is bounded above by
$$\frac{D}{K}(y^{w(I,i)}p^{n+aL})_{\tilde{a}} = \frac{D}{K}\pi(n+aL,\tilde{a},i)\,(r(I)p^{n+aL})_{\tilde{a}}. \qquad \square$$

The difference (†††) can be viewed as the sum of two terms: One term consists of the error resulting from the fact that, for each type (J, w) with $w = \tilde{a} - \delta_0(J)$, the number (∗∗) may not be integral. This term is bounded above (independently of n and $\tilde{a}$) by $|S|$. It will be clear that the estimates below easily accommodate this bounded amount. The second contribution to (†††) arises from
$$\{\gamma \in P'(n,a,i) : \tilde{a} \in \text{Log}(\text{mass}(\gamma)) \text{ and } \delta_0(\text{mass}(\gamma)) < \tilde{a}\}.$$
Since $\text{mass}(P'(n,a,i))_b = (y^{w(I,i)}p^{n+aL})_b$ for $b < \tilde{a}$, it follows from (7.9) that the absolute value of this contribution is less than
$$\frac{DM}{K}\pi(n+aL,\tilde{a},i)\,(r(I)p^{n+aL})_{\tilde{a}}.$$
Combining this with (††), our choice of K and the inequalities $D \geq 2$ and $|S| \leq M^2 D^2$, we find that $\widetilde{P}(n,a+1,i)$ contains the paths needed to repeat our correction procedure sufficiently many times to obtain sets $P(n,a+1,i)$ satisfying (†) for $b < \widehat{a+1}$. Actually, we have proved more: For $J \in S$ and an A_{F_0}-path η_0 such that $l(\eta_0) = L$ and $s(\eta_0) = J$, put
$$\begin{aligned} W(J,a,\eta_0) &= \{\gamma\eta_0 : \gamma \in W_{n+aL}(I,J,\tilde{a}-\delta_0(J))\}, \\ W(J,a,\eta_0,i) &= W(J,a,\eta_0) \cap P(n,a+1,i), \\ \nu(J,\tilde{a},\eta_0,i) &= |W(J,a,\eta_0,i)|/|W(J,a,\eta_0)|. \end{aligned}$$
The number of elements of $W(J,a,\eta_0)$ moved in (or out) of $\widetilde{P}(n,a+1,i)$ in making the corrections is bounded above by
$$\frac{M^2 D}{K}\pi(n+aL,\tilde{a},i)\,(r(I)p^{n+aL})_{\tilde{a}},$$
while (7.5) guarantees that $\widetilde{P}(n,a+1,i)$ contains at least
$$\frac{1}{2M^2|S|}\pi(n+aL,\tilde{a},i)\,(r(I)p^{n+aL})_{\tilde{a}}$$
elements of this set. The ratio of these two numbers does not exceed
$$\frac{2M^4 D|S|}{K} \leq \frac{2M^4 DM^2 D^2}{4^D M^{4D} D^{4D}} \leq \frac{2M^6 D^3}{16 M^8 D^8} \leq \frac{1}{8}.$$
Hence:

LEMMA 7.10. *If $a \leq \frac{3}{2}\epsilon_1 n$ and the set $W_{n+aL}(I,J,\tilde{a}-\delta_0(J))$ is nonempty, then for any A_{F_0}-path $\eta_0 \in W_L(J)$ we have*
$$|\nu(J,\tilde{a},\eta_0,i)/\pi(n+aL,\tilde{a},i) - 1| \leq \frac{1}{8}.$$

The next lemma will help us improve the inequality of (7.10) to the point where we can apply (7.8). Let $\bar{K} \geq 1$ be as determined by (7.6) for $\epsilon = \epsilon_0/(L+4)$. Fix $h_1 \in \mathbb{N}$ such that $h_1 \geq 10D$ and $2^{-h_1+1} \leq \xi/(4\bar{K})$.

LEMMA 7.11. *Let $a \in (\epsilon_1 n, \frac{3}{2}\epsilon_1 n)$ and $\theta \in (\frac{\xi}{4}, \frac{1}{8})$. Suppose that we have*
$$|\nu(J,\tilde{b},\eta_0,i)/\pi(n+bL,\tilde{b},i) - 1| \leq \theta$$
whenever $a - h_1 \leq b < a$, the set $W_{n+bL}(I, J, \tilde{b} - \delta_0(J))$ is nonempty and $\eta_0 \in W_L(J)$ is an A_{F_0}-path. Then

(i) *for each nonempty $W_{n+aL}(I, J, \tilde{a} - \delta_0(J))$ and each A_{F_0}-path $\eta_0 \in W_L(J)$ we have*
$$|\nu(J,\tilde{a},\eta_0,i)/\pi(n+aL,\tilde{a},i) - 1| \leq \theta/2,$$

(ii) *for $J \in S$ and $\tilde{a} - 4D \leq w < \tilde{a} - \delta_0(J)$ with $W_{n+aL}(I, J, w) \neq \emptyset$ we have*
$$\left| \frac{|W_{n+aL}(I,J,w) \cap P(n,a,i)|}{|W_{n+aL}(I,J,w)|} \cdot \frac{1}{\pi(n+aL, w+\delta_0(J), i)} - 1 \right| \leq 3\theta.$$

Before proving (7.11), we use it to complete our proof of (3.2) in the case of one variable. Let m be the smallest nonnegative integer such that $\frac{1}{8}2^{-m} \leq \xi/4$. Put $a_1 = \epsilon_1 n + mh_1$. Note that (7.10) and repeated application of (7.11)(i) show that
$$|\nu(J,\tilde{b},\eta_0,i)/\pi(n+bL,\tilde{b},i) - 1| \leq \xi/4$$
for $b \in [a_1 - h_1, a_1)$. Then (7.11)(ii) tells us that
$$\left| \frac{|W_{n+a_1L}(I,J,w) \cap P(n,a_1,i)|}{|W_{n+a_1L}(I,J,w)|} \cdot \frac{1}{\pi(n+a_1L, w+\delta_0(J), i)} - 1 \right| \leq \frac{3\xi}{4}$$
for $\tilde{a}_1 - 4D \leq w < \tilde{a}_1 - \delta_0(J)$. We shall apply (7.8) to paths of length $n + a_1 L$, with
$$a_0^* = (n + a_1 L)\delta_0 + a_1 - D/2.$$
Recall that ξ was determined by taking $\epsilon = \epsilon_0/(L+4)$ in (7.8). Note that for our choice of a_0^* we have
$$\frac{a_0^* - (n+a_1 L)\delta_0}{n + a_1 L} \in (\epsilon, 2\epsilon)$$
since the numerator differs from $\epsilon_1 n$ by a constant, the denominator differs from $n + \epsilon_1 nL$ by a constant, and our choices of $\epsilon_0 < 1/4$ and $\epsilon_1 = 2\epsilon = 2\epsilon_0/(L+4)$ place the ratio $\epsilon_1/(1+\epsilon_1 L)$ in the interval $(\epsilon, 2\epsilon)$. From $a = 0$ to $a = a_1$ we use the above inductive procedure, applying it simultaneously at the low end and the high end. This partitions
$$\{\gamma \in W_{n+a_1 L}(I) : \mathrm{mass}(\gamma) < \tilde{a}_1 \text{ or } \mathrm{mass}(\gamma) > \delta_\infty(I) + (n+a_1 L)\delta_\infty - a_1\}$$
into sets $P(n, a_1, i)$ such that
$$\mathrm{mass}(P(n,a_1,i))_b = (y^{w(I,i)} p^{n+a_1 L})_b$$

for $b < \tilde{a}_1 = \delta_0(I) + (n + a_1 L)\delta_0 + a_1$ as well as $b > \delta_\infty(I) + (n + a_1 L)\delta_\infty - a_1$. We then allocate each of the remaining sets $W_{n+a_1 L}(I, J, w)$ according to $\pi(n + a_1 L, w + \delta_0(J), i)$. Lemmas (7.4), (7.10) and (7.11) ensure that, provided n is large enough, (7.8) will apply and complete the proof of (3.2) in the case of one variable. It remains for us to prove (7.11).

PROOF OF (7.11). Letting $u \in [1, 5D]$, consider $J \in S$ such that $w = \tilde{a} - \delta_0(J) - u \in [\tilde{a} - 4D, \tilde{a} - \delta_0(J))$ and $W_{n+aL}(I, J, w) \neq \emptyset$. For $J_1 \in S$ and $h \in \mathbb{Z}^+$, denote by $\widetilde{W}_{hL}(J_1, J)$ the subset of $W_{hL}(J_1, J)$ consisting of those $\eta \in W_{hL}(J_1, J)$ such that $\delta_0(\mathrm{mass}(\eta)) = hL\delta_0 + \delta_0(J_1) + h - u + 1$ and, expressing $\eta = \eta_1 \eta_2 \cdots \eta_h$ in terms of subpaths η_j of length L, the inequality $\delta_0(\mathrm{mass}(\eta_1 \eta_2 \cdots \eta_j)) \leq jL\delta_0 + \delta_0(J_1) + j$ is valid for $j = 1, \ldots, h$. Observe that, since $|W_L(J_1)| \leq M^2 D^2$ for every $J_1 \in S$, we have $|\widetilde{W}_{hL}(J_1, J)| \leq M^{2h} D^{2h}$. It follows from the construction of $P(n, a, i)$ that $W_{n+aL}(I, J, w) \cap P(n, a, i)$ is a disjoint union

$$\bigcup_h \bigcup_{J_0} \bigcup_{\eta_0} W(J_0, a - h, \eta_0, i) \widetilde{W}_{(h-1)L}(t(\eta_0), J)$$

over $1 \leq h \leq a$, $J_0 \in S$ and A_{F_0}-paths $\eta_0 \in W_L(J_0)$: Here, we have written $W(J_0, a-h, \eta_0, i)\widetilde{W}_{(h-1)L}(t(\eta_0), J)$ for the set of all concatenations $\gamma \eta_0 \eta$ with $\gamma \eta_0 \in W(J, a-h, \eta_0, i)$ and $\eta \in \widetilde{W}_{(h-1)L}(t(\eta_0), J)$.

Consider h, J_0, η_0 such that $W(J_0, a-h, \eta_0)\widetilde{W}_{(h-1)L}(t(\eta_0), J)$ is nonempty. Suppose $h \geq h_1$. Write $J_1 = t(\eta_0)$, let $\eta \in \widetilde{W}_{(h-1)L}(J_1, J)$, and express $\eta = \eta_1 \eta_2 \cdots \eta_{h-1}$ in terms of subpaths η_j of length L. Since $\delta_0(\mathrm{mass}(\eta)) = (h-1)L\delta_0 + \delta_0(J_1) + h - u$,

$$\sum_{j=1}^{h-1} \mathrm{Log}(\mathrm{wt}(\eta_j)) + \delta_0(t(\eta_j)) - \delta_0(s(\eta_j)) = (h-1)L\delta_0 + h - u.$$

On the other hand, using the fact that $\mathrm{Log}(\mathrm{mass}(\eta_j)) \subset \mathrm{Log}(r(s(\eta_j))\, p^L)$,

$$0 \leq \mathrm{Log}(\mathrm{wt}(\eta_j)) + \delta_0(t(\eta_j)) - \delta_0(s(\eta_j)) - L\delta_0 \leq D.$$

It follows that at least $\frac{h-u}{D}$ of the η_j are not A_{F_0}-paths. Moreover, whenever η_j is not an A_{F_0}-path the quantity

$$\mathrm{Log}(\mathrm{wt}(\eta_j)) + \delta_0(t(\eta_j)) - \delta_0(s(\eta_j)) - L\delta_0$$
$$= \delta_0(\mathrm{mass}(\eta_j)) - \delta_0(r(s(\eta_j))\, p^L)$$

is a nonzero element of

$$\mathrm{Log}\left(r(s(\eta_j))\, p^L\right) - \delta_0\left(r(s(\eta_j))\, p^L\right) \subset \mathrm{Log}(p^{2L}) - 2L\delta_0,$$

allowing us to apply (7.1) to see that

$$\sum_{J_0}\sum_{\eta_0}\sum_{J}|W(J_0,a-h,\eta_0,i)\widetilde{W}_{(h-1)L}(t(\eta_0),J)|$$

$$\leq M^{2(h-1)}D^{2(h-1)}\sum_{J_0}\sum_{\eta_0}|W(J_0,a-h,\eta_0,i)|$$

$$\leq M^{2(h-1)}D^{2(h-1)}(y^{w(I,i)}p^{n+(a-h+1)L})_{\widetilde{a-h+1}-1}$$

$$\leq M^{2(h-1)}D^{2(h-1)}\left(\frac{1}{K}\right)^{\frac{h-u}{D}}(y^{w(I,i)}p^{n+aL})_{\tilde{a}-u}.$$

Now, using our requirement that $K > (4M^4D^4)^D$ we find that

$$(M^2D^2)^{u-1}\left(\frac{M^2D^2}{K^{\frac{1}{D}}}\right)^{h-u} \leq (M^2D^2)^{u-1}\left(\frac{1}{4M^2D^2}\right)^{h-u} \leq \left(\frac{1}{2}\right)^h,$$

where the second inequality holds because it is equivalent to

$$\left(\frac{1}{2}\right)^{h-u}\frac{(M^2D^2)^{u-1}}{(M^2D^2)^{h-u}} \leq \left(\frac{1}{2}\right)^u$$

and we have $h \geq h_1 \geq 10D$, $u \leq 5D$. Since $2^{-h_1+1} \leq \xi/(4\bar{K})$, it follows that

(‡) $\quad \sum_{h=h_1}^{a}\sum_{J_0}\sum_{\eta_0}\sum_J |W(J_0,a-h,\eta_0,i)\widetilde{W}_{(h-1)L}(t(\eta_0),J)|$

$$\leq \frac{\xi}{4\bar{K}}(y^{w(I,i)}p^{n+aL})_{\tilde{a}-u}.$$

Combining this with (7.4) and the fact that

(‡‡) $\quad |\nu(J_0,\widetilde{a-h},\eta_0,i)/\pi(n+(a-h)L,\widetilde{a-h},i) - 1| < \theta$

for $h < h_1$, we find that the number of corrections made in obtaining $P(n,a+1,i)$ from $\widetilde{P}(n,a+1,i)$ is no more than

$$(\xi/4+2\theta)M \sum_{\substack{u\in \text{Log}(p^L)-L\delta_0, \\ u\neq 0}} (y^{w(I,i)}p^{n+aL})_{\tilde{a}-u}.$$

This is exactly $(\xi/4+2\theta)$ times the amount we started lemma (7.9) with. With this factor, the argument following (7.9) proves (7.11)(i) since, instead of the bound $\frac{1}{8}$ appearing in (7.10), the argument yields the bound

$$\frac{1}{8}(\xi/4+2\theta) \leq \frac{1}{8}(3\theta) \leq \frac{\theta}{2}.$$

To verify (ii), first simplify notation by letting

$$W' = \bigcup_{h=1}^{h_1-1}\bigcup_{J_0}\bigcup_{\eta_0} W(J_0,a-h,\eta_0)\widetilde{W}_{(h-1)L}(t(\eta_0),J),$$

$$W'' = \bigcup_{h=h_1}^{a}\bigcup_{J_0}\bigcup_{\eta_0} W(J_0,a-h,\eta_0)\widetilde{W}_{(h-1)L}(t(\eta_0),J),$$

$$W = W_{n+aL}(I,J,w),$$

and $W'(i) = W' \cap P(n,a,i)$, $W''(i) = W'' \cap P(n,a,i)$, $W(i) = W \cap P(n,a,i)$. By the inequality (‡) and our choice of $\bar{K}$,

$$\begin{aligned}
|W''(i)| &\leq \frac{\xi}{4\bar{K}} (y^{w(I,i)} p^{n+aL})_{\tilde{a}-u} \\
&= \frac{\xi}{4\bar{K}} \pi(n+aL, \tilde{a}-u, i) \, (r(I) p^{n+aL})_{\tilde{a}-u} \\
&\leq \frac{\xi}{4} \pi(n+aL, \tilde{a}-u, i) \, |W|.
\end{aligned}$$

It follows that $|W''| \leq \frac{\xi}{4}|W|$ and, since $|W| = |W'| + |W''|$, that

$$\left(1 - \frac{\xi}{4}\right) |W| \leq |W'|.$$

Moreover, by (‡‡) and (7.4),

$$\left(1 - \frac{3}{2}\theta\right) \pi(n+aL, \tilde{a}-u, i) \, |W'| \leq |W'(i)| \leq \left(1 + \frac{3}{2}\theta\right) \pi(n+aL, \tilde{a}-u, i) \, |W'|$$

for large n. Using these inequalities and that $\frac{\xi}{4} \leq \theta$ and $w + \delta_0(J) = \tilde{a} - u$,

$$\begin{aligned}
\frac{|W(i)|}{|W|} \leq \frac{|W'(i)|}{|W'|} + \frac{|W''(i)|}{|W|} &\leq \left(1 + \frac{3}{2}\theta + \frac{\xi}{4}\right) \pi(n+aL, w+\delta_0(J), i) \\
&\leq (1 + 3\theta) \pi(n+aL, w+\delta_0(J), i)
\end{aligned}$$

and

$$\begin{aligned}
\frac{|W(i)|}{|W|} \geq \frac{|W'(i)|}{|W|} &\geq \left(1 - \frac{\xi}{4}\right) \frac{|W'(i)|}{|W'|} \\
&\geq \left(1 - \frac{\xi}{4}\right)\left(1 - \frac{3}{2}\theta\right) \pi(n+aL, w+\delta_0(J), i) \\
&\geq (1 - 3\theta)\pi(n+aL, w+\delta_0(J), i) \,,
\end{aligned}$$

which completes the proof of (7.11). □

8

Totally conforming eigenvectors for the general case

We turn to the general case of an irreducible R_k^+-matrix A with $\beta_A = p \in R_k^+$ and a right eigenvector r satisfying (a)–(c). In this section we show that for large n the matrix A^n may be state-split to produce a totally conforming right eigenvector. It should be clear that we are generalizing section 6 with the aid of an inductive procedure similar to one we employed in section 4.

PROPOSITION 8.1. *Let $p \in R_k^+$ and $K > 0$, and consider an extreme point e of $W(p)$. If $b \in \text{Log}(p^L) - Le$ for some $L \in \mathbb{N}$ and $b \neq 0$, then*

$$(p^n)_{ne+b} \geq K\, (p^n)_{ne}$$

for all large n.

Proposition (8.1) is a special case of (9.1), which is proved in [**MT4**]. The following lemma and its proof, which we omit, are similar to (4.6) and (6.4)

LEMMA 8.2. *Let F be a proper face of $W(A) = W(p)$, and let $m \geq 0$. There exists $N \in \mathbb{N}$ such that any $G(A)$-path γ with length $l(\gamma) > N$ and $\text{Log}(\text{wt}_A(\gamma)) \cap P(l(\gamma)F, m) \neq \emptyset$ must visit a state of an F-component of A.*

For $F \in \mathcal{F}(p)$, let S_F denote the union of the states of the F-components of A. Pick $L \in \mathbb{N}$ and, for $F \in \mathcal{F}(p)$ and $I \in S_F$, pick $a_{F,I} \in \text{Log}(r(I)_F)$ and $b_{F,I} \in \text{Log}(p_F^L)$ such that

$$\text{Log}(r(I)) + b_{F,I} - a_{F,I} \subset \text{Log}(p^L) \quad \text{and} \quad \text{Log}(r(I)_F) + b_{F,I} - a_{F,I} \subset \text{Log}(p_F^L).$$

Let

$$M = \max\{r(I)_w : I \in S, w \in \text{Log}(r(I))\}.$$

Fix D such that $D \geq \|b_{F,I}\|$ for all $b_{F,I}$ and $D \geq \|u\|$ for all $u \in \text{Log}(p^L) \cup \bigcup_{I \in S} \text{Log}(r(I))$. Apply (3.6) to find $H_F \geq H(p_F)$ and finite sets $T(F) \subset \mathbb{Z}^k$ such that, for all large n,

$$(*) \quad \begin{aligned} \text{Log}(p^n) &= \bigcup_{F \in \mathcal{F}(p)} (\text{Int}(p_F^n, H_F + 4D) + T(F)) \\ &= \bigcup_{F \in \mathcal{F}(p)} (\text{Int}(p_F^n, H_F) + T(F))\,. \end{aligned}$$

Fix m such that $m \geq H_{W(p)}$ and $m \geq 3D + \max\{\|w\| : w \in T(F), F \in \mathcal{F}(p)\}$. Putting $d = \dim(W(p))$ and $m_{d-1} = m$, choose $m_0 \geq m_1 \geq \cdots \geq m_{d-1}$ as in (4.5). For a proper face F of $W(p)$ write $m_F = m_{\dim(F)}$, and say that $\gamma \in W_n(I)$

53

is a *near-F path* if $\mathrm{Log}(\mathrm{mass}(\gamma))$ intersects $P(nF, m_F)$. Call $\gamma \in W_n(I)$ a *near-boundary* path if it is a near-F path for a proper face F of $W(p)$. Let N be large enough for (4.5) and (∗) to hold for all $n \geq N$ and for (8.2) to ensure that every near-F path of length N visits an F-component of A.

For each proper face F of $W(p)$, fix a vector v_F that exposes F. Recall that for $q \in R_k$ we are writing

$$\delta_v(q) = \min\{w \cdot v : w \in \mathrm{Log}(q)\}.$$

Consider an extreme point e of $W(p)$. If $q \in R_k$ $\{e\}$-conforms to p then $v = v_{\{e\}}$ exposes the single point $\mathrm{Log}(q_{\{e\}})$; in this case we put

$$\partial_e(q) = \mathrm{Log}(q_{\{e\}}).$$

As a result of (b), this notation applies to $r(I)$ for any $I \in S_e = S_{\{e\}}$, and to $\mathrm{mass}(\gamma)$ for any path γ with $t(\gamma) \in S_e$. For $i \in S_e$, we put $\partial_e(I) = \partial_e(r(I))$.

Write each state mass $r(I)$ as a sum of monomials in R_k:

$$r(I) = \sum_{i=1}^{\rho(I)} x^{w(I,i)}.$$

We will state-split large powers A^n of A by partitioning each $W_n(I)$ into sets (I, i), $1 \leq i \leq \rho(I)$. The sets (I, i) and their associated vectors $\alpha^{(I,i)}$ will satisfy the following requirements:

$$\mathrm{Log}(\alpha^{(I,i)} \cdot r) \subset \mathrm{Log}(x^{w(I,i)} p^n),$$

and, for each extreme point e of $W(p)$, the set (I, i) contains a path $\gamma^{(I,i,e)}$ with

$$\partial_e(\mathrm{mass}(\gamma^{(I,i,e)})) = ne + w(I, i).$$

Considering the matrix $\widehat{A^n}$ and the right eigenvector $\widehat{r}$ resulting from such a splitting, we have

$$\widehat{r}(I, i) = \alpha^{(I,i)} \cdot r = \sum_{\gamma \in (I,i)} \mathrm{mass}(\gamma),$$

and it is easy to see that the above requirements imply that every entry $\widehat{r}(I, i)$ is F-conforming for every proper face F of p^n

We start by establishing the existence of the desired paths $\gamma^{(I,i,e)}$. For $w \in \mathrm{Log}(r(I))$, observe that either $(p^n r(I))_{ne+w} \to \infty$ as $n \to \infty$, or the following applies: $p_e = 1$ and w is such that

$$(p^l r(I))_{le+w} = (p_e)^l r(I)_w = r(I)_w$$

for all l.

LEMMA 8.3. *Suppose that e is an extreme point of $W(p)$ and $w \in \mathrm{Log}(r(I))$ is such that $(p^n r(I))_{ne+w} \to \infty$ as $n \to \infty$. Then*

$$|\{\gamma \in W_n(I) : t(\gamma) \in S_e, \partial_e(\mathrm{mass}(\gamma)) = ne + w\}| \longrightarrow \infty$$

as $n \to \infty$.

8. TOTALLY CONFORMING EIGENVECTORS FOR THE GENERAL CASE

PROOF. Put $v = v_{\{e\}}$ and

$$\mu = \min\{l(\gamma)(e \cdot v) + w \cdot v - \delta_v(\mathrm{mass}(\gamma)) : s(\gamma) = I, l(\gamma)e + w \in \mathrm{Log}(\mathrm{mass}(\gamma))\}.$$

Let γ be a path such that $s(\gamma) = I$, $l(\gamma)e + w \in \mathrm{Log}(\mathrm{mass}(\gamma))$ and

$$l(\gamma)(e \cdot v) + w \cdot v - \delta_v(\mathrm{mass}(\gamma)) = \mu.$$

Write $J = t(\gamma)$. Observing that $l(\gamma)e + w - \mathrm{Log}(\mathrm{wt}(\gamma)) \in \mathrm{Log}(r(J))$, consider $\eta \in W_n(J)$ with $ne + l(\gamma)e + w - \mathrm{Log}(\mathrm{wt}(\gamma)) \in \mathrm{Log}(\mathrm{mass}(\eta))$. Recall that $\delta_v(\mathrm{mass}(\eta)) \geq \delta_v(p^n r(J))$, with equality if and only if η is an A_v-path. Since $(n + l(\gamma))e + w \in \mathrm{Log}(\mathrm{mass}(\gamma\eta))$ and

$$(n + l(\gamma))(e \cdot v) + w \cdot v - \delta_v(\mathrm{mass}(\gamma\eta))$$
$$= n(e \cdot v) + l(\gamma)(e \cdot v) + w \cdot v - \delta_v(\mathrm{mass}(\gamma)) + \delta_v(r(J)) - \delta_v(\mathrm{mass}(\eta))$$
$$= \mu + \delta_v(p^n r(J)) - \delta_v(\mathrm{mass}(\eta)),$$

it follows from the minimality of μ that η must be A_v-path. This shows that, by extending γ by a suitable A_v-path, we may assume $J \in S_e$. Assuming this, consider

$$a = l(\gamma)e + w - \mathrm{Log}(\mathrm{wt}(\gamma)) \in \mathrm{Log}(r(J))$$

By the above argument, any $\eta \in W_n(J)$ with $ne + a \in \mathrm{Log}(\mathrm{mass}(\eta))$ is an A_v-path. Hence,

$$(p^n)_{ne+a-\partial_e(J)} \, r(J)_{\partial_e(J)} \leq (p^n r(J))_{ne+a} \leq M \, (p^n r(J))_{ne+\partial_e(J)}.$$

Observing $M\,(p^n r(J))_{ne+\partial_e(J)} = M(p_e)^n\, r(J)_{\partial_e(J)}$, we find that

$$(p^n)_{ne+a-\partial_e(J)} \leq M(p^n)_{ne}.$$

Since $a - \partial_e(J) \in \mathrm{Log}(p^L) - Le$ by (b), this would contradict (8.1) unless $a = \partial_e(J)$. Therefore $\mu = 0$. Moreover, for any A_v-path $\eta \in W_n(J)$, the path $\gamma\eta$ has $\partial_e(\mathrm{mass}(\gamma\eta)) = (n + l(\gamma))e + w$. Consideration of those η that terminate in a principal $\{e\}$-component proves the lemma in the case $p_e > 1$.

Suppose $p_e = 1$. In this case, each $\{e\}$-component consists of a single periodic orbit. Since $(p^n r(I))_{ne+w} \to \infty$, we can find a path η such that $s(\eta) = I$, $ne + w \in \mathrm{Log}(\mathrm{mass}(\eta))$, $l(\eta) \geq N$ and, writing $\eta = \eta_1\eta_2$ with $l(\eta_1) = N$, the tail η_2 is not an A_v-path. By (8.2), the path η_1 must visit S_e. Let $I' = t(\eta)$ and $w' = ne + w - \mathrm{Log}(\mathrm{wt}(\eta)) \in \mathrm{Log}(r(I'))$ and apply the above argument to I' and w' to extend η to a path α with the following properties: $s(\alpha) = I$, $J = t(\alpha) \in S_e$, $l(\alpha)e + w = \partial_e(\mathrm{mass}(\alpha))$ and α can be decomposed $\alpha = \alpha_1\alpha_2$ so that $J_1 = t(\alpha_1) = s(\alpha_2) \in S_e$ and α_2 is not an A_v-path. Let $\tilde\alpha_1, \tilde\alpha_2$ be A_v-cycles of the same length l such that $s(\tilde\alpha_1) = t(\tilde\alpha_1) = J_1$ and $s(\tilde\alpha_2) = t(\tilde\alpha_2) = J$. For $j = 0, 1, \ldots, n$ the paths

$$\alpha_1\,(\tilde\alpha_1)^j\,\alpha_2\,(\tilde\alpha_2)^{n-j}$$

provide $n + 1$ distinct elements of

$$\{\gamma \in W_{l(\alpha)+nl}(I) : t(\gamma) \in S_e,\, \partial_e(\mathrm{mass}(\gamma)) = ne + w\}. \qquad \square$$

For $F \in \mathcal{F}(p)$, let W_F be the set of all paths γ_0 such that $l(\gamma_0) \leq N$, $s(\gamma_0) = I$, $t(\gamma_0) \in S_F$ and γ_0 does not visit S_F prior to its terminal state. Using our choice of N and considering first entry into S_F, we see that every near-F path γ is of the form $\gamma = \gamma_0 \gamma_1$ for some $\gamma_0 \in W_F$. (Note that, in the case $I \in S_F$, the set W_F consists of the empty path, so that γ_0 is the empty path and $\gamma_1 = \gamma$.) Except for minor notational changes, the proof of the following lemma is the same as that of (6.6).

LEMMA 8.4. *Let e be an extreme point of $W(p)$. Suppose $p_e = 1$ and $w \in \mathrm{Log}(r(I))$ is such that $(p^l \, r(I))_{le+w} = (p_e)^l \, r(I)_w = r(I)_w$ for all l. Then*

$$|\{\gamma \in W_n(I) : t(\gamma) \in S_e \,,\, \partial_e(\mathrm{mass}(\gamma)) = ne + w\}| = r(I)_w \,.$$

Whenever $w = w(I,i)$ is such that $(p^n r(I))_{ne+w} \to \infty$ as $n \to \infty$, the existence of the desired paths $\gamma^{(I,i,e)}$ is guaranteed by (8.3) for all large n. The condition $(p^n r(I))_{ne+w} \to \infty$ fails only when (8.4) applies to provide exactly $r(I)_w$ paths $\gamma \in W_n(I)$ such that $t(\gamma) \in S_e$ and $\partial_e(\mathrm{mass}(\gamma)) = ne + w$; we bijectively assign these paths to the $r(I)_w$ sets (I,i) with $w(I,i) = w$.

For every extreme point e of $W(p)$ and $1 \leq i \leq \rho(I)$, we have thus assigned to (I,i) a path $\gamma^{(I,i,e)}$ with $\partial_e(\mathrm{mass}(\gamma^{(I,i,e)})) = ne+w(I,i)$ and $\mathrm{Log}\left(\mathrm{mass}(\gamma^{(I,i,e)})\right) \subset \mathrm{Log}\left(x^{w(I,i)} p^n\right)$. The remaining near-boundary paths will be allocated via an induction on the affine dimension of $F \in \mathcal{F}(p)$. Suppose F is a proper face of $W(p)$ and all near-G paths have been allocated for all $G \in \mathcal{F}(p)$ with $\dim(G) < \dim(F)$. Let $\gamma \in W_n(I)$ be a near-F path which is not a near-G path for any $G \in \mathcal{F}(p)$ with $\dim(G) < \dim(F)$. Write $\gamma = \gamma_0 \gamma_1$ with $\gamma_0 \in W_F$. Let $l = l(\gamma_0)$ and $J = t(\gamma_0)$. Since

$$\mathrm{Log}(\mathrm{mass}(\gamma_0)) \in \mathrm{Log}(p^l r(I)),$$

we can find $i \in \{1, \ldots, \rho(I)\}$ and $a \in \mathrm{Log}(p^l)$ such that

$$\mathrm{Log}(\mathrm{wt}(\gamma_0)) + a_{F,J} = w(I,i) + a.$$

We then have

$$\begin{aligned}
\mathrm{Log}(\mathrm{mass}(\gamma)) &\subset \mathrm{Log}(\mathrm{wt}(\gamma_0) r(J) p^{n-l}) \\
&\subset \mathrm{Log}(\mathrm{wt}(\gamma_0) p^L p^{n-l}) + a_{F,J} - b_{F,J} \\
&= \mathrm{Log}(p^{n+L-l}) + w(I,i) + a - b_{F,J} \\
&\subset \mathrm{Log}(p^{n+L}) + w(I,i) - b_{F,J} \,.
\end{aligned}$$

We claim that $\mathrm{Log}(\mathrm{mass}(\gamma)) \subset \mathrm{Log}(p^n) + w(I,i)$ for large n. To verify this, let $a \in \mathrm{Log}(\mathrm{mass}(\gamma))$. Use the previous inclusion and $(*)$ to find $G \in \mathcal{F}(p)$ and $w \in T(G)$ such that

$$u + b_{F,J} - w(I,i) \in \mathrm{Int}(p_G^{n+L}, H_G + 4D) + w \,.$$

By our choices of D and $m \leq m_G$, the path γ is then near-G. Using (4.5) it follows, as in the proof of (4.7), that $F \subset G$. In particular,

$$b_{F,J} \in \mathrm{Log}(p_F^L) \subset \mathrm{Log}(p_G^L) \subset \mathrm{Log}\left(c_{p_G}^L \Delta(p_G)\right).$$

8. TOTALLY CONFORMING EIGENVECTORS FOR THE GENERAL CASE

We continue to argue as in the proof of (4.7): Express $u + b_{F,J} - w(I,i) - w = v + \tilde{v}$ with $v \in \text{Log}(p_G^n)$ and $\tilde{v} \in \text{Log}(p_G^L)$. If $v' \in \partial W(p_G^n)$ then, finding $v'' \in \partial W(p_G^L)$ such that $v' + v'' \in \partial W(p_G^{n+L})$, we see that

$$\begin{aligned} \|v - v'\| &= \|v + \tilde{v} - (v' + v'') - \tilde{v} + v''\| \\ &\geq H_G + 4D - (\|\tilde{v}\| + \|v''\|) \\ &\geq H_G + 2D. \end{aligned}$$

That is,
$$v \in \text{Int}(p_G^n, H_G + 2D).$$

Since $\tilde{v} - b_{F,J} \in \text{Log}(\Delta(p_G))$ and $\|\tilde{v}\|, \|b_{F,J}\| \leq D$, it follows that

$$u - w(I,i) - w = v + \tilde{v} - b_{F,J} \in \text{Int}(p_G^n, H_G)$$

and, considering (∗), that

$$u - w(I,i) \in \text{Int}(p_G^n, H_G) + w \subset \text{Log}(p^n).$$

Thus $\text{Log}(\text{mass}(\gamma)) \subset \text{Log}(x^{w(I,i)} p^n)$ as claimed. We assign γ to the set (I,i). provided γ is not an element of $\{\gamma^{(I,j,e)} : 1 \leq j \leq \rho(I), e \text{ extreme point of } W(p)\}$.

Having dealt with the allocation of near-boundary paths, suppose $\gamma \in W_n(I)$ is not a near-boundary path. Since $\text{Log}(\text{mass}(\gamma))) \subset \text{Log}(p^n r(I))$, we can find $u \in \text{Log}(\text{mass}(\gamma))$ and $i \in \{1, \ldots, \rho(I)\}$ such that

$$u - w(I,i) \in \text{Log}(p^n).$$

As γ is not near-F for any proper face $F \in \mathcal{F}(p)$ and $T(W(p)) = \{0\}$, we see from (∗) that

$$u - w(I,i) \in \text{Int}(p^n, H_{W(p)} + 4D).$$

By (3.4)(i), this implies

$$\text{Log}(\text{mass}(\gamma)) - w(I,i) \subset \text{Log}(\Delta(p)).$$

By (∗) and our choice of D we then have

$$\text{Log}(\text{mass}(\gamma)) - w(I,i) \subset \text{Int}(p^n, H_{W(p)} + 2D) \subset \text{Log}(p^n),$$

and we assign γ to the set (I,i).

This completes the state-splitting. Repeating verbatim (not even notational changes are required) the proof of (6.7), we obtain:

PROPOSITION 8.5. *For n large enough, A^n may be state-split in such a way that the resulting matrix $\widehat{A^n}$ has a right eigenvector $\hat{r}$ which totally conforms to p^n.*

9

Splitting the conforming eigenvector in the general case

In this section we complete the proof of (3.2) by generalizing the splitting procedure of section 7.

Let A be an irreducible R_k^+-matrix with $\beta_A = p \in R_k^+$ and a right eigenvector r satisfying (a)–(c). We assume A is over $R(\Delta(A))^+$ and $p \in R(\Delta(p))^+$. We also assume that $\text{Log}(p)$ generates $\mathbb{Z}^k$; that is, $R(\Delta(p)) = R_k$. Replacing A by a suitable power, (5.3) and (8.5) enable us to also assume without loss of generality that r totally conforms to p and that every F-component of A is aperiodic for every $F \in \mathcal{F}(p)$.

For each proper face F of $W(p)$, we fix $v_F \in \mathbb{Z}^k$ which exposes F. We make sure that each v_F has coprime entries. It follows from the total conformity of r to p that the faces A_{v_F} and r_{v_F} are independent of the choice of the vector exposing F; we denote them by A_F and r_F. We put $\delta_F = \delta_{v_F}(p)$, $\delta_F(I) = \delta_{v_F}(r(I))$ for $I \in S$, and $\delta_F(\text{mass}(\gamma)) = \delta_{v_F}(\text{mass}(\gamma))$ for A-paths γ. In this section S_F will denote the union of the states of the *principal F-components* of A. Picking a representative of the coset $c_{p_F} \Delta(p_F)$, we will write $c_F = \text{Log}(c_{p_F})$.

Recall that ∂F is the boundary of $F \in \mathcal{F}(p)$ taken in the affine hull of F. That is, ∂F is the union of $G \in \mathcal{F}(p)$ that are proper subsets of F. The set $B(\partial F, \epsilon)$ is the ϵ-neighbourhood of ∂F in $\mathbb{Q}^k$. We will make use of the following three results from [**MT4**]. They are general versions of (7.1)–(7.3).

PROPOSITION 9.1. *Let F be a proper face of $W(p)$. Let $L \in \mathbb{N}$ and $K, \bar{\epsilon}_F > 0$. There exist $\epsilon_F > 0$ and $N \in \mathbb{N}$ such that for $a \in \mathbb{Z}^k$ the inequality*

$$(p^n)_{a+b} \geq K(p^n)_a$$

holds whenever $n \geq N$, $\frac{a}{n} \in B(F, \epsilon_F) \backslash B(\partial F, \bar{\epsilon}_F)$ and $b \in (\text{Log}(p^L) \backslash \text{Log}(p_F^L)) - \text{Log}(p_F^L)$.

PROPOSITION 9.2. *Let $F \in \mathcal{F}(p)$. Let $\bar{\epsilon}_F, \theta > 0$ and $D \in \mathbb{N}$. There exist $\epsilon_F > 0$ and $N \in \mathbb{N}$ such that*

$$\left| \frac{(p^n)_{a+b-w}}{(p^n)_{a+b-w'}} \frac{(p^n)_{a+b'-w'}}{(p^n)_{a+b'-w}} - 1 \right| < \theta$$

whenever $n \geq N$ and $a, b, b', w, w' \in \mathbb{Z}^k$ are such that $(p^n)_{a+b-w}, (p^n)_{a+b-w'} \neq 0$, $\frac{a}{n} \in B(F, \epsilon_F) \backslash B(\partial F, \bar{\epsilon}_F)$, $b - b' \in \text{Log}(\Delta(p_F))$ and $\|b\|, \|b'\|, \|w\|, \|w'\| \leq D$.

PROPOSITION 9.3. *Let* $F \in \mathcal{F}(p)$. *Let* $\bar{\epsilon}_F > 0$, $l \in \mathbb{Z}$ *and* $D \in \mathbb{N}$. *There exist* $\epsilon_F > 0$ *and* $K, N \in \mathbb{N}$ *such that*

$$\frac{1}{K}(p^{n+l})_{a+lc_F+b} \leq (p^n)_a \leq K(p^{n+l})_{a+lc_F+b}$$

whenever $n \geq N$ *and* $a, b \in \mathbb{Z}^k$ *are such that* $\frac{a}{n} \in B(F, \epsilon_F) \setminus B(\partial F, \bar{\epsilon}_F)$, $b \in \text{Log}(\Delta(p_F))$ *and* $\|b\| \leq D$.

Fix $I \in S$. As before, write $r(I) = \sum_{i=1}^{\rho(I)} x^{w(I,i)}$. For $n \in \mathbb{N}$ and $a \in \text{Log}(r(I)p^n)$, let $\pi(n, a)$ be the probability vector of length $\rho(I)$ whose i-th entry equals

$$\pi(n, a, i) = (x^{w(I,i)} p^n)_a / (r(I) p^n)_a.$$

The following is obtained from (9.2) in the way that (7.4) was obtained from (7.2).

LEMMA 9.4. *Let* $\theta, \bar{\epsilon}_F > 0$ *and* $D, l \in \mathbb{Z}^+$. *There exist* $\epsilon_F > 0$ *and* $N \in \mathbb{N}$ *such that*

$$|\pi(n+l, a+lc_F+b, i)/\pi(n, a, i) - 1| < \theta$$

whenever $n \geq N$, $\frac{a}{n} \in B(F, \epsilon_F) \setminus B(\partial F, \bar{\epsilon}_F)$, $(p^n)_a \neq 0$, $b \in \text{Log}(\Delta(p_F))$ *and* $\|b\| \leq D$.

Using (a)–(c) and the fact that r totally conforms to p, pick $L \in \mathbb{Z}^+$, $u_{F,J} \in \mathbb{Z}^k$ and, for each principal F-component $\mathcal{C}$ and coset $x^u \Delta(A_{\mathcal{C}}) \in \Delta(p_F)/\Delta(A_{\mathcal{C}})$, pick $c_{\mathcal{C}} \in \Delta(p_F)$, $D_{F,J} \in \text{Log}(\Delta(p_F))$ and polynomials $q^{(F,u)}(J) \in R(\Delta(A_{\mathcal{C}}))$ such that:

(1) $\text{Log}(r(J)) + u_{F,J} \subset \text{Log}(p^L)$ and $\text{Log}(r_F(J)) + u_{F,J} \subset \text{Log}(p_F^L)$ for every $J \in S$,
(2) for each principal F-component $\mathcal{C}$, the matrix $\tilde{A}_{\mathcal{C}}$ indexed by (the states of) $\mathcal{C}$ and with

$$\tilde{A}_{\mathcal{C}}(J, J') = A_{\mathcal{C}}(J, J') \frac{x^{u_{F,J}} x^{D_{F,J}}}{c_{\mathcal{C}} c_{p_F} x^{u_{F,J'}} x^{D_{F,J'}}}$$

is over $R(\Delta(A_{\mathcal{C}}))$,
(3) $\sum_{J \in \mathcal{C}} q^{(F,u)}(J) x^{u_{F,J}} x^{D_{F,J}} r_F(J) = x^u p_F^L$ for every $F \in \mathcal{F}(p)$, principal F-component $\mathcal{C}$ and $x^u \Delta(A_{\mathcal{C}}) \in \Delta(p_F)/\Delta(A_{\mathcal{C}})$,
(4) if $F \in \mathcal{F}(p)$ and $\mathcal{C}$ is a principal F-component then, for any $J, J' \in \mathcal{C}$, there is an A_F-path of length $L - |S|$ from J to J'.

For $F \in \mathcal{F}(p)$, we let

$$h_F = \text{l.c.m.}\{|\Delta(p_F)/\Delta(A_{\mathcal{C}})| : \mathcal{C} \text{ principal } F\text{-component of } A\}.$$

Observe that if $x^w \in \Delta(p_F)$ then $x^{h_F w} \in \Delta(A_{\mathcal{C}})$ for every principal F-component $\mathcal{C}$ and that, for every $A_{\mathcal{C}}$-cycle γ with $h_F | l(\gamma)$, we have

$$\text{wt}_A(\gamma) \in c_{p_F}^{l(\gamma)} \Delta(A_{\mathcal{C}})$$

as a result of (2). This leads us to also make sure that

(5) L is divisible by h_F for every $F \in \mathcal{F}(p)$.

9. SPLITTING THE EIGENVECTOR IN THE GENERAL CASE

Proceeding as in section 7, we fix an even number D such that $D \geq 2\|u\|$ for the representative x^u of each coset appearing in (3); $D \geq \|v_F\|$, $D > 2\|u_{F,J}\|$, $D > 2\|D_{F,J}\|$ for every $F \in \mathcal{F}(p)$ and $J \in S$; and $D > 2\|w\|$ whenever

$$w \in \text{Log}(p^L) \cup \bigcup_{J \in S} \text{Log}(r(J)) \cup \bigcup_{F,J,u} \text{Log}(q^{(F,u)}(J)).$$

We let

$$M = \max\{(p^L)_w, r(J)_w, |q^{(F,u)}(J)_w| : w \in \mathbb{Z}^k\}.$$

We define $W_n(I, J)$, $W_n(I, J, w)$ and mass(W) of $W \subset W_n(I)$ as in section 7, and say that $\gamma \in W_n(I, J, w)$ is of *type* (n, J, w). The paths provided by the next proposition will enable us to use (3) to make corrections.

PROPOSITION 9.5. *Let* $F \in \mathcal{F}(p)$, $\bar{\epsilon}_F > 0$ *and* $\bar{D} \in \mathbb{N}$. *There exist* $\epsilon_F > 0$ *and* $N, \bar{K} \in \mathbb{N}$ *such that, for every* $n \geq N$ *and* $a \in \mathbb{Z}^k$ *with* $\frac{a}{n} \in B(F, \epsilon_F) \setminus B(\partial F, \bar{\epsilon}_F)$, *we can find a principal F-component* $\mathcal{C}$, $J_1 \in \mathcal{C}$ *and* $x^u \Delta(A_\mathcal{C}) \in \Delta(p_F)/\Delta(A_\mathcal{C})$ *with*

$$|W_n(I, J_1, a - u + b - Lc_F + u_{F,J_1} + D_{F,J_1})| \geq \frac{1}{\bar{K}}(r(I)p^n)_a$$

whenever $b \in \text{Log}(\Delta(A_\mathcal{C}))$ *and* $\|b\| \leq \bar{D}$.

The following lemma, which we need for the proof of (9.5), is similar to (7.7).

LEMMA 9.6. *There exist* $l_1 \in \mathbb{N}$ *and* $w_1 \in \mathbb{Z}^k$ *so that the following are valid for all* $F \in \mathcal{F}(p)$.

  (i) l_1 *is divisible by* h_F.
  (ii) *For every principal F-component* $\mathcal{C}$, *we have* $x^{w_1} \in c_{p_F}^{l_1} \Delta(A_\mathcal{C})$.
  (iii) *For every principal F-component* $\mathcal{C}$, $J \in \mathcal{C}$ *and* $h \in \mathbb{N}$, *the set*

$$\{x^{hw_1 + b} : b \in \text{Log}(\Delta(A_\mathcal{C})), \|b\| \leq h\}$$

  is contained in $\{\text{wt}(\gamma) : \gamma \in W_{hl_1}(J, J)\}$.

PROOF. By (6.1) of [**MT1**] and (5) of [**T3**], each $\Delta(A_\mathcal{C})$ has the same rank as $\Delta(p_F)$, namely, the affine dimension $d = \dim(F)$. Find $l \in \mathbb{N}$ and, for each principal F-component $\mathcal{C}$, paths $\gamma_{\mathcal{C},j}$, $\eta_{\mathcal{C},j}$, $1 \leq j \leq d$, such that $l(\gamma_{\mathcal{C},j}) = l(\eta_{\mathcal{C},j}) = l$, each $\gamma_{\mathcal{C},j}$ and $\eta_{\mathcal{C},j}$ visits every state of $\mathcal{C}$ and $\{\text{wt}(\gamma_{\mathcal{C},j})/\text{wt}(\eta_{\mathcal{C},j}) : 1 \leq j \leq d\}$ generates $\Delta(A_\mathcal{C})$. Note that

$$\text{wt}(\gamma_{\mathcal{C},j})^{h_F}, \text{wt}(\eta_{\mathcal{C},j})^{h_F} \in c_{p_F}^{lh_F} \Delta(A_{\mathcal{C}_0})$$

for all principal F-components $\mathcal{C}, \mathcal{C}_0$. Fixing a principal F-component $\mathcal{C}_0$, use (14) of [**MT2**] to find $l_0 \in \mathbb{N}$ and $A_\mathcal{C}$-cycles $\tau_\mathcal{C}$ such that $l(\tau_\mathcal{C}) = l_0 h_F$ and

$$\text{wt}(\tau_\mathcal{C}) \prod_{j=1}^{d} \text{wt}(\gamma_{\mathcal{C},j})^{h_F} \text{wt}(\eta_{\mathcal{C},j})^{h_F} = \text{wt}(\tau_{\mathcal{C}_0}) \prod_{j=1}^{d} \text{wt}(\gamma_{\mathcal{C}_0,j})^{h_F} \text{wt}(\eta_{\mathcal{C}_0,j})^{h_F}.$$

Let

$$w_1 = \text{Log}\left(\text{wt}(\tau_{\mathcal{C}_0})^k \prod_{j=1}^{d} \text{wt}(\gamma_{\mathcal{C}_0,j})^{k h_F} \text{wt}(\eta_{\mathcal{C}_0,j})^{k h_F}\right)$$

and $l_1 = (l_0 + 2ld)kh_F$. For any principal F-component $\mathcal{C}$, we then have $x^{w_1} \in c_{p_F}^{l_1} \Delta(A_\mathcal{C})$, and the desired $A_\mathcal{C}$-cycles are among

$$\tau_\mathcal{C}^{kh} (\gamma_{\mathcal{C},1}^{khh_F - k_1} \eta_{\mathcal{C},1}^{khh_F + k_1})(\gamma_{\mathcal{C},2}^{khh_F - k_2} \eta_{\mathcal{C},2}^{khh_F + k_2}) \cdots (\gamma_{\mathcal{C},d}^{khh_F - k_d} \eta_{\mathcal{C},d}^{khh_F + k_d}),$$

$k_1, \ldots, k_d \in \mathbb{Z}$, $|k_1|, \ldots, |k_d| \leq kh$. □

PROOF OF (9.5). For each $J_0 \in S$, we use (4) to find a principal F-component $\mathcal{C}_{J_0}$ and, for every $J \in \mathcal{C}_{J_0}$, an A_F-path $\eta^{J_0 J} \in W_L(J_0, J)$. We shall apply (9.6) with $h = 4D + \bar{D}$. Let $l = L + l_1 h$. Consider

$$M'_n(a) = \{\gamma \in W_n(I) : a \in \text{Log}(\text{mass}(\gamma)_F)\},$$
$$M''_n(a) = \{\gamma \in W_n(I) : a \in \text{Log}(\text{mass}(\gamma)) \setminus \text{Log}(\text{mass}(\gamma)_F)\},$$

and apply (9.1) to find $0 < \epsilon_F < \bar{\epsilon}_F$ such that

$$\frac{1}{2}(r(I)p^{n-l})_{a - hw_1 - Lc_F} \leq \text{mass}(M'_{n-l}(a - hw_1 - Lc_F))_{a - hw_1 - Lc_F}$$

for all large n and $a \in B(F, \epsilon_F) \setminus B(\partial F, \bar{\epsilon}_F)$. For such n and a we can then find $J_0 \in S$ and $a_0 \in \text{Log}(r_F(J_0))$ with

$$|W_{n-l}(I, J_0, a - hw_1 - Lc_F - a_0)| \geq \frac{1}{2MD|S|}(r(I)p^{n-l})_{a - hw_1 - Lc_F}.$$

Consider $\mathcal{C} = \mathcal{C}_{J_0}$ and any $J_1 \in \mathcal{C}$. Since $\eta^{J_0 J_1}$ is an A_F-path, the equation $A_F^L r_F = p_F^L r_F$ provides $b' \in \text{Log}(p_F^L)$, $a'_0 \in \text{Log}(r_F(J_0))$, $a'_1 \in \text{Log}(r_F(J_1))$ with

$$b' + a'_0 = \text{Log}(\text{wt}(\eta^{J_0 J_1})) + a'_1.$$

Observe that, as a result of (1),

$$u' = a_0 - a'_0 - b' + a'_1 + u_{F,J_1} + D_{F,J_1}$$
$$= (a_0 - a'_0) - (b' - c_F^L) + (a'_1 + u_{F,J_1} - c_F^L) + D_{F,J_1} \in \Delta(p_F).$$

Let $x^u \Delta(A_\mathcal{C}) \in \Delta(p_F)/\Delta(A_\mathcal{C})$ be such that $u' - u \in \text{Log}(\Delta(A_\mathcal{C}))$. Noting that $\|u' - u\| \leq 4D$, for any $x^b \in \Delta(A_\mathcal{C})$ with $\|b\| \leq \bar{D}$ we can apply (9.6) to find $\eta \in W_{hl_1}(J_1, J_1)$ such that

$$\text{Log}(\text{wt}(\eta)) = hw_1 + u' - u + b.$$

For any $\gamma \in W_{n-l}(I, J_0, a - hw_1 - Lc_F - a_0)$, the path $\gamma \eta^{J_0 J_1} \eta \in W_n(I, J_1)$ then has

$$\text{Log}(\text{wt}(\gamma \eta^{J_0 J_1} \eta)) = a - Lc_F - u + b + u_{F,J_1} + D_{F,J_1}.$$

Hence,

$$|W_n(I, J_1, a - u + b - Lc_F + u_{F,J_1} + D_{F,J_1})| \geq \frac{1}{2MD|S|}(r(I)p^{n-l})_{a - hw_1 - Lc_F}.$$

Finally, since

$$a - hw_1 - Lc_F = a - lc_F + (l_1 hc_F - hw_1)$$

and $x^{l_1 hc_F - hw_1} \in \Delta(A_\mathcal{C})$, we can employ (9.3) to see that $(r(I)p^{n-l})_{a - hw_1 - Lc_F}$ is greater than a constant multiple of $(r(I)p^n)_a$. □

Recall that our goal is to partition $W_n(I)$ into sets (I, i), $1 \leq i \leq \rho(I)$, so that
$$\text{mass}((I, i)) = x^{w(I,i)} p^n.$$
We assume $\rho(I) \geq 2$, for otherwise we would have nothing to do. For $j \in \{0, 1, \ldots, k\}$, let
$$\mathcal{F}_j(p) = \{F \in \mathcal{F}(p) : \dim(F) = j\}.$$
For each $F \in \mathcal{F}(p)$ and $J \in S$, pick $a_{F,J} \in \text{Log}(r_F(J))$. Put $a_J = a_{W(p),J}$. It will be convenient to use v_F to specify neighbourhoods of nF in $\text{Log}(r(I)p^n)$: For $F \in \mathcal{F}(p)$ and $\lambda \in \mathbb{Z}$, define
$$\mathcal{N}_n(F, \lambda) = \{a \in \text{Log}(r(I)p^n) : a \cdot v_F \leq \delta_F(I) + n\delta_F + \lambda\}$$
and
$$\mathcal{N}_n(\partial F, \lambda) = \bigcup_{\substack{G \in \mathcal{F}(p), \\ G \subset F, G \neq F}} \mathcal{N}_n(G, \lambda).$$
For $j \in \{0, 1, \ldots, k-1\}$, write
$$\mathcal{N}_n(j, \lambda) = \bigcup_{F \in \mathcal{F}_j(p)} \mathcal{N}_n(F, \lambda)$$
and
$$\mathcal{N}_n(j, \lambda_0, \ldots, \lambda_j) = \bigcup_{j'=0}^{j} \mathcal{N}_n(j', \lambda_{j'}).$$

Fix $\epsilon^* > 0$ such that, for $F, G \in \mathcal{F}(p)$, we have $\mathcal{N}_n(F, 2\epsilon^* n) \cap \mathcal{N}_n(G, 2\epsilon^* n) \neq \emptyset$ if and only if $F \cap G \neq \emptyset$.

Let $W_n(F, \lambda)$, $W_n(\partial F, \lambda)$, $W_n(j, \lambda)$, $W_n(j, \lambda_0, \ldots, \lambda_j)$ respectively denote the union of the sets $W_n(I, J, w)$ over all types (n, J, w) such that $w + \text{Log}(r(J))$ intersects $\mathcal{N}_n(F, \lambda)$, $\mathcal{N}_n(\partial F, \lambda)$, $\mathcal{N}_n(j, \lambda)$, $\mathcal{N}_n(j, \lambda_0, \ldots, \lambda_j)$. We will sometimes drop the subscript from our notation for these objects if the length is clear from the context.

Consider $\xi, \bar{D} \geq 0$ and a partition of $W_n(k-1, \lambda_0, \ldots, \lambda_{k-1})$ into sets $P(n, i)$ satisfying two conditions:

(i) For every $a \in \mathcal{N}_n(k-1, \lambda_0, \ldots, \lambda_{k-1})$ we have
$$\text{mass}(P(n, i))_a = (x^{w(I,i)} p^n)_a.$$

(ii) For every type (n, J, w) such that
$$W_n(I, J, w) \subset W_n(k-1, \lambda_0, \ldots, \lambda_{k-1}) \setminus W_n(k-1, \lambda_0 - \bar{D}, \ldots, \lambda_{k-1} - \bar{D}),$$
we have
$$1 - \xi \leq \frac{|W_n(I, J, w) \cap P(n, i)|}{|W_n(I, J, w)|} \cdot \frac{1}{\pi(n, w + a_J, i)} \leq 1 + \xi.$$

Picking a suitable $\bar{D}$ and arguing as in the proof of (7.8), we obtain the following lemma. (Here, it suffices to take $\bar{D} = 3D^2$.)

9. SPLITTING THE EIGENVECTOR IN THE GENERAL CASE

LEMMA 9.7. *For every $\epsilon \in (0, \epsilon^*)$ there exist $\xi \geq 0$ and $N \in \mathbb{N}$ such that, for any $n \geq N$ and $\lambda_0, \ldots, \lambda_{k-1} \in (n\epsilon, n\epsilon^*)$, the existence of a partition of $W_n(k-1, \lambda_0, \ldots, \lambda_{k-1})$ into $\rho(I)$ sets $P(n, i)$ satisfying (i) and (ii) implies the existence of a partition of $W_{n+L}(I)$ into $\rho(I)$ sets (I, i) such that $\mathrm{mass}((I, i)) = x^{w(I,i)} p^{n+L}$.*

Let $d = 3(k-1)D^2$, $\widetilde{d} = (k-1)d + 1$ and $\widetilde{L} = \widetilde{d}L$. We will also have a finite sequence of numbers
$$\epsilon^* > \epsilon_0 > \epsilon_1 > \cdots > \epsilon_{k-1} > 0$$
whose selection we postpone until after we have described their use in the construction of a partition to which (9.7) applies. Let us write $\epsilon(0) = 0$ and $\epsilon(j) = \epsilon_0 + \epsilon_1 + \cdots + \epsilon_{j-1}$ for $j \in \{1, \ldots, k\}$.

When dealing with $r(I)p^l$ for some length l, we will refer to
$$\{a \in \mathrm{Log}(r(I)p^l) : a \cdot v_F - \delta_F(r(I)p^l) = \lambda\}$$
as the *F-level λ exponents*, to
$$\{\gamma \in W_l(I) : \delta_F(\mathrm{mass}(\gamma)) - \delta_F(r(I)p^l) = \lambda\}$$
as the *F-level λ paths*, and to the union of these objects over $F \in \mathcal{F}_j(p)$ as the *j-level λ exponents and paths*. Our construction will start with paths of length n and involve $\epsilon(k)n$ stages which we count with $\lambda \in \{1, \ldots, \epsilon(k)n\}$. At each stage the lengths of the paths are increased by $\widetilde{L}$. Starting with $\lambda = 1$, at each stage we include a new 0-level among those partitioned. No higher dimensional levels are allocated until $\lambda = \epsilon(1)n + 1$. From $\lambda = \epsilon(1)n + 1$ we allocate, in addition to a new 0-level, a new 1-level at every stage. In general, we start at $\lambda = \epsilon(j)n + 1$ to include j-levels among those allocated. Each stage of the construction consists of $\widetilde{d}$ steps, the lengths of the paths being increased by L at each step. Each of the $\widetilde{d}$ steps has a specific role. Numbering the steps $1, \ldots, \widetilde{d}$ and putting $d_j = (k-1-j)d + 1$, steps $\mu \in \{d_{j+1} + 1, \ldots, d_j\}$ are concerned with j-levels: If λ is such that no new j-level is to be introduced at stage λ, that is, if $\lambda \leq \epsilon(j)n$, step $(\lambda - 1)\widetilde{d} + \mu$ does nothing but extend paths by L. For $\lambda \geq \epsilon(j)n + 1$ the new j-level $\lambda - \epsilon(j)n - 1$ is allocated at step $(\lambda - 1)\widetilde{d} + d_j$. If no higher dimensional level than j is allocated at stage λ (that is, if $\lambda \leq \epsilon(j+1)n$), for each $\mu \in \{d_{j+1} + 1, \ldots, d_j - 1\}$ step $(\lambda - 1)\widetilde{d} + \mu$ just extends paths by L. Otherwise, the $d-1$ j-levels immediately preceding the new j-level may have been perturbed by steps concerned with higher dimensional levels; for $\mu \in \{d_{j+1} + 1, \ldots, d_j - 1\}$ step $(\lambda - 1)\widetilde{d} + \mu$ is dedicated to the corrections to j-level $\lambda - \epsilon(j)n + \mu - d_j - 1$. (Note that in the case $j = k-1$ only step $d_{k-1} = 1$ is concerned with $(k-1)$-levels and is dedicated to allocating the new $(k-1)$-level, if any; there are no higher dimensional levels to cause perturbations.) We will make this precise by spelling out the sets $P(n, \lambda, i, \mu)$, $1 \leq i \leq \rho(I)$, we have at the end of step $(\lambda - 1)\widetilde{d} + \mu$ and the sets $P(n, \lambda, i) = P(n, \lambda, i, \widetilde{d})$ we have at the end of stage λ.

Let $l(n, \lambda, \mu) = n + (\lambda - 1)\widetilde{L} + \mu L$ and $l(n, \lambda) = n + \lambda \widetilde{L}$. For $m \in \{0, 1, \ldots, k-1\}$ and $\mu \in \{d_{m+1}, d_{m+1} + 1, \ldots, d_m - 1\}$, let $\mathcal{N}_1(n, (\lambda - 1)\widetilde{d} + \mu)$ denote the set
$$\mathcal{N}_{l(n,\lambda,\mu)}(k-1, \lambda - 2, \ldots, \lambda - \epsilon(m)n - 2, \lambda - \epsilon(m+1)n - 1, \ldots, \lambda - \epsilon(k-1)n - 1).$$

This set represents the exponents considered through step $(\lambda - 1)\tilde{d} + \mu$, using the length in operation at the end of the step. To capture the corresponding paths, let $W_1(n, (\lambda - 1)\tilde{d} + \mu)$ denote the set

$$W_{l(n,\lambda,\mu)}(k-1, \lambda-2, \ldots, \lambda - \epsilon(m)n - 2, \lambda - \epsilon(m+1)n - 1, \ldots, \lambda - \epsilon(k-1)n - 1).$$

Let $P(n, 0, i) = \emptyset$. To continue inductively, suppose $\epsilon(j)n < \lambda \leq \epsilon(j+1)n$ and we have partitioned

$$W_1(n, (\lambda - 1)\tilde{d}) = W_{l(n,\lambda-1)}(j, \lambda - 2, \lambda - \epsilon(1)n - 2, \ldots, \lambda - \epsilon(j)n - 2)$$

into $\rho(I)$ sets $P(n, \lambda - 1, i)$ with

$$\mathrm{mass}(P(n, \lambda - 1, i))_a = (x^{w(I,i)} p^{l(n,\lambda-1)})_a$$

for all $a \in \mathcal{N}_1(n, (\lambda - 1)\tilde{d})$. Let

$$P(n, \lambda, i, 0) = P(n, \lambda - 1, i).$$

For $\mu \in \{1, \ldots, d_j - 1\}$ define

$$P(n, \lambda, i, \mu) = \{\gamma\eta : \gamma \in P(n, \lambda, i, \mu - 1), \eta \in W_L(t(\gamma)), \gamma\eta \in W_1(n, (\lambda - 1)\tilde{d} + \mu)\}.$$

If $\mu = d_m$ for some $m \in \{0, 1, \ldots, j\}$, first allocate according to $\pi(l(n, \lambda, d_m - 1), w + a_{F,J})$ every $W_{l(n,\lambda,d_m-1)}(I, J, w)$ that is contained in

$$W_{l(n,\lambda,d_m-1)}(F, \lambda - \epsilon(m)n - 1) \setminus W_1(n, (\lambda - 1)\tilde{d} + d_m - 1)$$

for some $F \in \mathcal{F}_m(p)$. Let $P'(n, \lambda, i, d_m - 1)$ be the set containing these allocations and $P(n, \lambda, i, d_m - 1)$. Then define $\tilde{P}(n, \lambda, i, d_m)$ to be the set

$$\{\gamma\eta : \gamma \in P'(n, \lambda, i, d_m - 1), \eta \in W_L(t(\gamma)), \gamma\eta \in W_1(n, (\lambda - 1)\tilde{d} + d_m)\}.$$

The sets $P(n, \lambda, i, d_m)$ will be obtained by using (3) to make corrections by exchanging paths among $\tilde{P}(n, \lambda, i, d_m)$, $1 \leq i \leq \rho(I)$. (The correction process will be explained below.) This will ensure that

$$\mathrm{mass}(P(n, \lambda, i, d_m))_a = (x^{w(I,i)} p^{l(n,\lambda,d_m)})_a$$

for all $a \in \mathcal{N}_{l(n,\lambda,d_m)}(j, \lambda - d, \ldots, \lambda - \epsilon(m-1)n - d, \lambda - \epsilon(m)n - 1, \ldots, \lambda - \epsilon(j)n - 1)$.

If $d_{m+1} < \mu < d_m$ for some $m \in \{0, 1, \ldots, j-1\}$, define $\tilde{P}(n, \lambda, i, \mu)$ to be the set

$$\{\gamma\eta : \gamma \in P(n, \lambda, i, \mu - 1), \eta \in W_L(t(\gamma)), \gamma\eta \in W_1(n, (\lambda - 1)\tilde{d} + \mu)\}.$$

The sets $P(n, \lambda, i, \mu)$ will be obtained from $\tilde{P}(n, \lambda, i, \mu)$ by using (3) to make corrections to m-level $\lambda - \epsilon(m)n + \mu - d_m - 1$. We will then have

$$\mathrm{mass}(P(n, \lambda, i, \mu))_a = (x^{w(I,i)} p^{l(n,\lambda,\mu)})_a$$

for all $a \in \mathcal{N}_{l(n,\lambda,\mu)}(j, \lambda - d, \ldots, \lambda - \epsilon(m-1)n - d, \lambda - \epsilon(m)n + \mu - d_m - 1, \lambda - \epsilon(m+1)n - 1, \ldots, \lambda - \epsilon(j)n - 1)$.

How are corrections made? Letting $\epsilon(j)n < \lambda \leq \epsilon(j+1)n$ as before, first consider the case $\mu = d_m$ for some $m \in \{0, 1, \ldots, j\}$. It is easy to see that

$$\mathrm{mass}(P'(n, \lambda, i, d_m - 1))_{a'} = (x^{w(I,i)} p^{l(n,\lambda,d_m-1)})_{a'}$$

for $a' \in \mathcal{N}_{l(n,\lambda,d_m-1)}(j, \lambda - d, \ldots, \lambda - \epsilon(m-1)n - d, \lambda - \epsilon(m)n - 2, \lambda - \epsilon(m+1)n - 1, \ldots, \lambda - \epsilon(j)n - 1)$. Hence, for $F \in \mathcal{F}_m(p)$, in the difference

$$x^{w(I,i)} p^{l(n,\lambda,d_m)} - \mathrm{mass}(\widetilde{P}(n,\lambda,i,d_m))$$

all terms in F-levels lower than $\lambda - \epsilon(m)n - 1$ cancel, and the F-level $\lambda - \epsilon(m)n - 1$ terms are given by

$$(x^{w(I,i)} p^{l(n,\lambda,d_m-1)} - \mathrm{mass}(P'(n,\lambda,i,d_m-1)))_F \, p_F^L.$$

It follows that the desired corrections amount to the addition of

$$(x^{w(I,i)} p^{l(n,\lambda,d_m-1)})_{a'} - \mathrm{mass}(P'(n,\lambda,i,d_m-1))_{a'}$$

copies of $x^{a'} p_F^L$ to $\mathrm{mass}(\widetilde{P}(n,\lambda,i,d_m))$ for each

$$a' \in \mathcal{N}_{l(n,\lambda,d_m-1)}(F, \lambda - \epsilon(m)n - 1) \setminus \mathcal{N}_{l(n,\lambda,d_m-1)}(\partial F, \lambda - \epsilon(m-1)n - d).$$

Since $\lambda - \epsilon(m-1)n - d \geq \epsilon_{m-1}n - d$, (9.5) may be applied to such a'. (Considering the fact that $\lambda - \epsilon(m)n \leq (\epsilon_m + \cdots + \epsilon_{k-1})n \leq k\epsilon_m n$, we will make sure $k\epsilon_m \leq \epsilon_F$ for $F \in \mathcal{F}_m(p)$.) By (9.5), we have $\bar{K} = \bar{K}_F$, a principal F-component $\mathcal{C}$, $J_1 \in \mathcal{C}$ and $x^u \Delta(A_\mathcal{C}) \in \Delta(p_F)/\Delta(A_\mathcal{C})$ such that

$$(*) \qquad |W_{l(n,\lambda,d_m-1)}(I, J_1, a' - u + b' - Lc_F + u_{F,J_1} + D_{F,J_1})|$$
$$\geq \tfrac{1}{K_F}(r(I) p^{l(n,\lambda,d_m-1)})_{a'}$$

as long as $b' \in \mathrm{Log}(\Delta(A_\mathcal{C}))$ and $\|b'\| \leq \bar{D}$. (The size of $\bar{D}$ will soon be clear.) From (3) we obtain the equation

$$\sum_{J \in \mathcal{C}} \sum_{\substack{b \in \mathrm{Log}(\Delta(A_\mathcal{C})), \\ \|b\| < D/2}} x^{a' - u + b + u_{F,J} + D_{F,J}} q^{(F,u)}(J)_b \, r_F(J) = x^{a'} p_F^L.$$

This means that if, for each $J \in \mathcal{C}$ and $b \in \mathrm{Log}(\Delta(A_\mathcal{C}))$ with $\|b\| < D/2$, we move $q^{(F,u)}(J)_b$ elements of $W_{l(n,\lambda,d_m)}(I, J, a' - u + b + u_{F,J} + D_{F,J})$ from $\widetilde{P}(n,\lambda,i',d_m)$ to $\widetilde{P}(n,\lambda,i,d_m)$ we will have increased $\mathrm{mass}(\widetilde{P}(n,\lambda,i,d_m))$ by $x^{a'} p_F^L$ and decreased $\mathrm{mass}(\widetilde{P}(n,\lambda,i',d_m))$ by $x^{a'} p_F^L$. Using (4) to fix $A_\mathcal{C}$-paths $\eta^{J_1,J} \in W_L(J_1, J)$ for $J \in \mathcal{C}$, we have

$$b' = b - u_{F,J_1} - D_{F,J_1} - \mathrm{Log}(\mathrm{wt}(\eta^{J_1 J})) + Lc_F + u_{F,J} + D_{F,J} \in \mathrm{Log}(\Delta(A_\mathcal{C})).$$

With this choice of b', $(*)$ provides paths $\gamma \in W_{l(n,\lambda,d_m-1)}(I, J_1)$ such that

$$\gamma \eta^{J_1 J} \in W_{l(n,\lambda,d_m)}(I, J, a' - u + b + u_{F,J} + D_{F,J}).$$

Our task below will be to verify that the ϵ_j can be chosen so that the paths supplied by $(*)$ are sufficient in number to make the required corrections.

At this point observe that the paths $\eta = \gamma \eta^{J_1 J}$ exchanged in the above transfer of $x^{a'} p_F^L$ do not perturb $\mathcal{N}_{l(n,\lambda,d_m)}(G, \lambda - \epsilon(\dim(G))n - 1)$ for $G \in \mathcal{F}(p)$ with $\dim(G) > m$. However, for $G \in \mathcal{F}_{m-1}(p)$ it is possible for $\mathrm{Log}(\mathrm{mass}(\eta))$ to intersect $\mathcal{N}_{l(n,\lambda,d_m)}(G, \lambda - \epsilon(m-1)n - 2)$: Recall that $a' \cdot v_G \geq \lambda - \epsilon(m-1)n - 1$. It is easy to see from the above that any $b \in \mathrm{Log}(\mathrm{mass}(\eta)) \cap \mathcal{N}_{l(n,\lambda,d_m)}(G, \lambda - \epsilon(m-1)n - 2)$ will have distance less than $3D$ from a'. Since $\|v_G\| \leq D$, it follows that

$$b \cdot v_G \geq \lambda - \epsilon(m-1)n - 3D^2.$$

In other words, the corrections made at $\mu = d_m$ can perturb only the last $3D^2 - 1$ of the $(m-1)$-levels and, by the same argument, at most the last $6D^2 - 1$ of the $(m-2)$-levels, and so on. In the extreme case, corrections to a $(k-1)$-level perturb at most the last $3(k-1)D^2 - 1$ of the 0-levels; this is why we took $d = 3(k-1)D^2$. These perturbations are compensated for in the following way.

In the case $\mu \in \{d_{m+1} + 1, \ldots d_m - 1\}$ for some $m \in \{0, 1, \ldots, j-1\}$, we have

$$\mathrm{mass}(\widetilde{P}(n, \lambda, i, \mu))_a = (x^{w(I,i)} p^{l(n,\lambda,\mu)})_a$$

for $a \in \mathcal{N}_{l(n,\lambda,\mu)}(j, \lambda - d, \ldots, \lambda - \epsilon(m-1)n - d, \lambda - \epsilon(m)n + \mu - d_m - 2, \lambda - \epsilon(m+1)n - 1, \ldots, \lambda - \epsilon(j)n - 1)$. In order to correct F-level $\lambda - \epsilon(m)n + \mu - d_m - 1$ for $F \in \mathcal{F}_m(p)$, we need to add to $\mathrm{mass}(\widetilde{P}(n, i, \lambda, \mu))$ the quantity

$$\sum_{a'} \left(x^{w(I,i)} p^{l(n,\lambda,\mu-1)} - \mathrm{mass}\left(P(n, \lambda, i, \mu-1)\right) \right)_{a'} x^{a'} p_F^L,$$

where the sum is over a' such that

$$a' \cdot v_F = \delta_F(I) + l(n, \lambda, \mu - 1)\delta_F + \lambda - \epsilon(m)n + \mu - d_m - 1$$

and

$$(x^{w(I,i)} p^{l(n,\lambda,\mu-1)})_{a'} - \mathrm{mass}\left(P(n, \lambda, i, \mu-1)\right)_{a'} \neq 0.$$

In transferring $x^{a'} p_F^L$ for such a', we will look back to step $(\lambda + \mu - d_m - 1)\widetilde{d} + d_m$, during which F-level $\lambda - \epsilon(m)n + \mu - d_m - 1$ was allocated. For states J_1, J that belong to the same principal F-component, fix an A_F-path

$$\eta^{J_1 J} \in W_{(d_m - \mu)\widetilde{L} + (\mu - d_m + 1)L}(J_1, J).$$

Note that

$$l(n, \lambda, \mu - 1) - l(n, \lambda + \mu - d_m, d_m - 1) = (d_m - \mu)(\widetilde{L} - L).$$

Since $(r(I) p^{l(n,\lambda,\mu-1)})_{a'} \neq 0$ and $a' \in \mathcal{N}_{l(n,\lambda,\mu-1)}(F, k\epsilon_m n) \setminus \mathcal{N}_{l(n,\lambda,\mu-1)}(\partial F, \frac{\epsilon_{m-1}}{2} n)$, we will have

$$(r(I) p^{l(n,\lambda+\mu-d_m, d_m-1)})_{a' - (d_m - \mu)(\widetilde{L} - L) c_F} \neq 0.$$

So, (9.5) provides a principal F-component $\mathcal{C}$, $J_1 \in \mathcal{C}$, $x^u \Delta(A_\mathcal{C}) \in \Delta(p_F)/\Delta(A_\mathcal{C})$ and a supply of elements of

$$W_{l(n,\lambda+\mu-d_m,d_m-1)}(I, J_1, a' - (d_m - \mu)(\widetilde{L} - L)c_F - u + b' - Lc_F + u_{F,J_1} + D_{F,J_1}),$$

as long as $b' \in \mathrm{Log}(\Delta(A_\mathcal{C}))$ and $\|b'\| \leq \overline{D}$. Consider (3) for this choice of $\mathcal{C}$, u, and let $b \in \mathrm{Log}(q^{(F,u)}(J))$ for some $J \in \mathcal{C}$. Observing that

$$b' = b - u_{F,J_1} - D_{F,J_1} - \mathrm{Log}(\mathrm{wt}(\eta^{J_1 J})) + (d_m - \mu)(\widetilde{L} - L)c_F + Lc_F + u_{F,J} + D_{F,J}$$

is an element of $\mathrm{Log}(\Delta(A_\mathcal{C}))$, we obtain from above a supply of paths γ such that

$$\gamma \eta^{J_1 J} \in W_{l(n,\lambda,\mu)}(I, J, a' - u + b + u_{F,J} + D_{F,J}).$$

These paths and (3) enable us to transfer $x^{a'} p_F^L$ between $\widetilde{P}(n, \lambda, i, \mu)$, $1 \leq i \leq \rho(I)$. This completes the description of the corrections that produce the sets $P(n, \lambda, i, \mu)$ from the sets $\widetilde{P}(n, \lambda, i, \mu)$, subject to verifying that the ϵ_j may be chosen to guarantee that the supply of paths is rich enough to make all of the required corrections.

By decreasing ϵ^* if necessary, make sure $0 < k\epsilon^* \widetilde{L} < \frac{1}{2}$. We will inductively choose $1 \leq K_0 < K_1 < \cdots < K_{k-1}$, $1 \leq \bar{K}_0 < \bar{K}_1 < \cdots < \bar{K}_{k-1}$ and $\theta_0 > \theta_1 > \cdots > \theta_{k-1} > 0$ along with $\epsilon^* > \epsilon_0 > \epsilon_1 > \cdots > \epsilon_{k-1} > 0$. For each $j \in \{0, 1, \ldots, k-1\}$ the constants K_j, $\bar{K}_j$ and θ_j will have the property that (9.1) holds with $K = K_j$, (9.5) holds with $\bar{K} = \bar{K}_j$ and (9.4) holds with $\theta = \theta_j$ on $\mathcal{N}_n(F, k\epsilon_j n) \setminus \mathcal{N}_n(\partial F, \frac{\epsilon_{j-1}}{3} n)$ for all $F \in \mathcal{F}_j(p)$ and all large n. We will first pick ϵ_j, $\bar{K}_j$ by applying (9.5) to each $F \in \mathcal{F}_j(p)$. Then, by decreasing ϵ_j while keeping $\bar{K}_j$ constant, we will be able to pick a sufficiently large $K_j > 4M^4 D^{4k} \bar{K}_j$ and sufficiently small θ_j to guarantee the supply of paths for our corrections. We preface the selection of ϵ_j, K_j, $\bar{K}_j$, θ_j with some notation and related observations.

Let $j \in \{0, 1, \ldots, k-1\}$. Suppose that, instead of starting to allocate higher dimensional levels from stage $\epsilon(j+1)n+1$, we continued to adhere to the rules in effect between $\epsilon(j)n+1$ and $\epsilon(j+1)n$. Let, for $1 \leq \lambda \leq \epsilon(j)n + (k-j)\epsilon_j n$ and $1 \leq \mu \leq \widetilde{d}$, the resulting sets be denoted by $P^{(j)}(n, \lambda, i, \mu)$. Also write $P^{(j)}(n, \lambda, i, 0) = P^{(j)}(n, \lambda-1, i, \widetilde{d})$. Put

$$\mathcal{N}_1^{(j)}(n, (\lambda-1)\widetilde{d}+\mu) = \mathcal{N}_{l(n,\lambda,\mu)}(j, \lambda-2, \lambda-\epsilon(1)n-2, \ldots, \lambda-\epsilon(j)n-2)$$

for $1 \leq \mu < d_j$, and

$$\mathcal{N}_1^{(j)}(n, (\lambda-1)\widetilde{d}+\mu)$$
$$= \mathcal{N}_{l(n,\lambda,\mu)}(j, \lambda-2, \ldots, \lambda-\epsilon(m-1)n-2, \lambda-\epsilon(m)n-1, \ldots, \lambda-\epsilon(j)n-1)$$

when $d_m \leq \mu < d_{m-1}$ for $m \leq j$. Let $W_1^{(j)}(n, (\lambda-1)\widetilde{d}+\mu)$ be the union of $W_{l(n,\lambda,\mu)}(I, J, w)$ such that $w + \mathrm{Log}(r(J))$ intersects $\mathcal{N}_1^{(j)}(n, (\lambda-1)\widetilde{d}+\mu)$. The sets $P^{(j)}(n, \lambda, i, \mu)$ form a partition of $W_1^{(j)}(n, (\lambda-1)\widetilde{d}+\mu)$. Also note that we have $P^{(j)}(n, \lambda, i, \mu) = P(n, \lambda, i, \mu)$ for $(\lambda-1)\widetilde{d}+\mu \leq \epsilon(j+1)n\widetilde{d}$. In particular $P^{(k-1)}(n, \lambda, i, \mu)$ are identical to $P(n, \lambda, i, \mu)$.

Let $m \leq j$ and $\bar{D} \in \mathbb{N}$. Suppose $h \in \mathbb{N}$ and $W_{l(n,\lambda,d_m-1)}(I, J, w)$ is contained in $W_{l(n,\lambda,d_m-1)}(m, \lambda-\epsilon(m)n-1)$ but not in

$$W_{l(n,\lambda,d_m-1)}(j, \lambda-2, \ldots, \lambda-\epsilon(m)n-2, \lambda-\epsilon(m+1)n-1, \ldots, \lambda-\epsilon(j)n-1).$$

Consider $\eta \in W_{(h\widetilde{d}+d_j-d_m)L}(J)$ such that, writing $\eta = \eta_1 \eta_2 \cdots \eta_{h\widetilde{d}+d_j-d_m}$ in terms of subpaths of length L, we have

$$W_{l(n,\lambda,d_m-1)}(I, J, w)\eta_1 \cdots \eta_{h'} \subset W_1^{(j)}(n, (\lambda-1)\widetilde{d}+d_m-1+h')$$

for $h' = 1, 2, \ldots, h\widetilde{d}+d_j-d_m$. Let $W^{(j)}(\lambda, m, J, w, \eta, i)$ denote the set of $\gamma \in W_{l(n,\lambda,d_m-1)}(I, J, w)$ such that $\gamma\eta \in P^{(j)}(n, \lambda+h, i, d_j-1)$. Put

$$\nu^{(j)}(\lambda, m, J, w, \eta, i) = |W^{(j)}(\lambda, m, J, w, \eta, i)|/|W_{l(n,\lambda,d_m-1)}(I, J, w)|.$$

Observe that these objects stabilize from $h = d$: If $h \geq d$ then, decomposing $\eta = \eta'\eta''$ with $l(\eta') = d\widetilde{d}+d_j-d_m$, we have

$$W^{(j)}(\lambda, m, J, w, \eta, i) = W^{(j)}(\lambda, m, J, w, \eta', i).$$

In the case $j = k-1$ we will often omit the superscript $(k-1)$ from our notation.

For $W_{l(n,\lambda,d_j-1)}(I,J,w)$ contained in $\mathcal{N}_1^{(j)}(n,(\lambda-1)\tilde{d}+d_j-1)$, we can consider all allowed $W^{(j)}(\lambda-h,m,J_0,w_0,\eta,i)$ with $\eta \in W_{(h\tilde{d}+d_j-d_m)L}(J_0,J)$ and $w = w_0 + \text{Log}(\text{wt}(\eta))$ to express $W_{l(n,\lambda,d_j-1)}(I,J,w) \cap P^{(j)}(n,\lambda,i,d_j-1)$ as a disjoint union

$$(**) \qquad \bigcup_{h\in\mathbb{N}} \bigcup_{m=0}^{j} \bigcup_{\substack{J_0,w_0,\eta, \\ w=w_0+\text{Log}(\text{wt}(\eta))}} W^{(j)}(\lambda-h,m,J_0,w_0,\eta,i)\eta.$$

This decomposition will be put to use before long. Similarly for $\bar{D} \in \mathbb{N}$ and

$$a \in \mathcal{N}_1^{(j)}(n,(\lambda-1)\tilde{d}+d_j-1) \setminus \mathcal{N}_{l(n,\lambda,d_j-1)}(j,\lambda-2-\bar{D},\ldots,\lambda-\epsilon(j)n-2-\bar{D}),$$

we can consider all allowed $W^{(j)}(\lambda-h,m,J_0,w_0,\eta,i)$ with $\eta \in W_{(h\tilde{d}+d_j-d_m)L}(J_0)$ and $a \in w_0 + \text{Log}(\text{mass}(\eta))$ to express $\{\gamma \in P^{(j)}(n,\lambda,i,d_j-1) : a \in \text{Log}(\text{mass}(\gamma))\}$ as a disjoint union

$$\bigcup_{h\in\mathbb{N}} \bigcup_{m=0}^{j} \bigcup_{\substack{J_0,w_0,\eta, \\ a\in w_0+\text{Log}(\text{mass}(\eta))}} W^{(j)}(\lambda-h,m,J_0,w_0,\eta,i)\eta.$$

Considering the terms corresponding to a fixed value of $h \in \mathbb{N}$:

LEMMA 9.8. *Let $\bar{D} \in \mathbb{N}$. There exist $C > 0$ and $\alpha \in (0,1)$ such that for all*

$$a \in \mathcal{N}_1^{(j)}(n,(\lambda-1)\tilde{d}+d_j-1) \setminus \mathcal{N}_{l(n,\eta,d_j-1)}(j,\lambda-2-\bar{D},\ldots,\lambda-\epsilon(j)n-2-\bar{D}),$$

every $h \in \mathbb{N}$ and large n we have

$$\sum_{m=0}^{j} \sum_{\substack{J_0,w_0,\eta, \\ a\in w_0+\text{Log}(\text{mass}(\eta))}} |W^{(j)}(\lambda-h,m,J_0,w_0,\eta,i)| \leq C\alpha^h \left(x^{w(I,i)} p^{l(n,\lambda,d_j-1)}\right)_a.$$

(9.8) is verified by reasoning as in the first part of the proof of (7.11). The basic idea is that for every $W^{(j)}(\lambda-h,m,J_0,w_0,\eta,i)$ appearing in the above sum we can find $F \in \mathcal{F}(p)$ with $\dim(F) \leq j$ and

$$a \cdot v_F - (w_0 \cdot v_F + \delta_F(J_0)) \geq h - \bar{D}.$$

For large h, this allows for manifold applications of (9.1). (The number of applications is proportional to h; the constant C takes care of small h.) Using the fact that $K_m > K_0 \geq 4M^4 D^{4k}$, one then obtains (9.8).

We turn to the inductive selection of $\epsilon_j, K_j, \bar{K}_j, \theta_j$. To start the process, let $0 < \epsilon_0 < \epsilon^*$, $\bar{K}_0$ and $K_0 = 4M^4 D^{4k}\bar{K}_0$ be such that (9.5) holds with $\bar{K} = \bar{K}_0$ and (9.1) holds with $K = K_0$ on $\mathcal{N}_n(F, 2k\epsilon_0 n)$ for all $F \in \mathcal{F}_0(p)$ and large n. By shrinking $\epsilon_0 > 0$ (while keeping $\bar{K}_0, K_0$ constant) we will be able to also ensure that (9.4) holds for arbitrarily small $\theta_0 > 0$ on $\mathcal{N}_n(0, 2k\epsilon_0 n)$ for all large n.

For $\lambda \leq \epsilon(1)n$, the corrections require, for each i,

$$(x^{w(I,i)} p^{l(n,\lambda,d_0-1)})_{a'} - \text{mass}(P'(n,\lambda,i,d_0-1))_{a'}$$

exchanges of $x^{a'}p_F^L$ for suitable $F \in \mathcal{F}_0(p)$. Since $|\mathrm{Log}(r(J))| < D^k$ for all $J \in S$, it follows from (9.1) that the absolute value of the above difference is less than

$$\frac{MD^k}{K_0}\pi(l(n,\lambda,d_0-1),a',i)\left(r(I)p^{l(n,\lambda,d_0-1)}\right)_{a'}.$$

So, for each type used in exchanging $x^{a'}p_F^L$, we need no more than M times the above number of paths of the type in $\widetilde{P}(n,\lambda,i,d_0-i)$. Meanwhile (9.5) guarantees at least

$$\frac{1}{\bar{K}_0}\pi(l(n,\lambda,d_0-1),a',i)\left(r(I)p^{l(n,\lambda,d_0-1)}\right)_{a'}$$

paths of the type. Consideration of the ratio of these numbers yields the analogue of (7.10) for $j = 0$. In fact, working with $P^{(0)}(n,\lambda,i,\mu)$ instead of $P(n,\lambda,i,\mu)$, we see that the sets $P^{(0)}(n,\lambda,i,\mu)$ may be constructed for $1 \leq \lambda \leq k\epsilon_0 n$, and that the following is valid: If $1 \leq \lambda \leq k\epsilon_0 n$ and $W_{l(n,\lambda,d_0-1)}(I,J,w)$ is contained in $W_{l(n,\lambda,d_0-1)}(F,\lambda-1)\setminus W_{l(n,\lambda,d_0-1)}(0,\lambda-2)$ for some $F \in \mathcal{F}_0(p)$, then for every $W^{(0)}(\lambda,0,J,w,\eta,i)$ we have

$$|\nu^{(0)}(\lambda,0,J,w,\eta,i)/\pi(l(n,\lambda,d_0-1),w+a_{F,J},i)-1| < \frac{1}{8}.$$

This kindles the following generalization of (7.11).

LEMMA 9.9. *Let* $j \in \{0,1,\ldots,k-1\}$. *The constants* $\epsilon_m, K_m, \bar{K}_m, \theta_m, 0 \leq m \leq j$, *may be chosen to have the property that for every* $\xi > 0$ *there exists* $N \in \mathbb{N}$ *such that the following are valid for* $n \geq N$.

(i) *It is possible to construct* $P^{(j)}(n,\lambda,i,\mu)$ *for* $1 \leq \lambda \leq \epsilon(j)n + (k-j)\epsilon_j n$.
(ii) *If* $\epsilon(j)n + \frac{\epsilon_j}{3}n \leq \lambda \leq \epsilon(j)n + (k-j)\epsilon_j n$, $m \in \{0,1,\ldots,j\}$ *and the set* $W_{l(n,\lambda,d_m-1)}(I,J,w)$ *is contained in* $W_{l(n,\lambda,d_m-1)}(F,\lambda-\epsilon(m)n-1)$ *for some* $F \in \mathcal{F}_m(p)$ *but not in*

$$W_{l(n,\lambda,d_m-1)}(j,\lambda-2,\ldots,\lambda-\epsilon(m)n-2,\lambda-\epsilon(m+1)n-1,\ldots,\lambda-\epsilon(j)n-1),$$

then for every $W^{(j)}(\lambda,m,J,w,\eta,i)$ *we have*

$$|\nu(\lambda,m,J,w,\eta,i)/\pi(l(n,\lambda,d_m-1),w+a_{F,J},i)-1| < \xi.$$

Before proving (9.9) we use it to complete our proof of (3.2). Let $\epsilon_m, K_m, \bar{K}_m, \theta_m, 0 \leq m \leq k-1$ be as provided by (9.9) in the case $j = k-1$. Then let $\xi > 0$ be as given by (9.7) for $\epsilon = \epsilon_{k-1}/3$. Consider $P(n,\epsilon(k-1)n+\frac{2}{3}\epsilon_{k-1}n,i)$. Observe that, since $\epsilon^* > \epsilon_0 > \cdots > \epsilon_{k-1} > 0$ and $k\epsilon^*\widetilde{L} < \frac{1}{2}$, we have

$$\frac{\frac{2}{3}\epsilon_{k-1}n}{n+\epsilon(k-1)n\widetilde{L}+\frac{3}{4}\epsilon_{k-1}n\widetilde{L}} > \frac{\frac{1}{2}\epsilon_{k-1}n}{n+\epsilon(k)n\widetilde{L}} > \frac{\epsilon_{k-1}}{2(1+k\epsilon^*\widetilde{L})} > \frac{\epsilon_{k-1}}{3}$$

Write $\lambda = \epsilon(k-1)n + \frac{2}{3}\epsilon_{k-1}n$. Applying (9.4) and (9.5) with $F = W(p)$ and $\bar{\epsilon}_{W(p)} = \epsilon_{k-1}/3$, we see that the inequalities in (9.4) and (9.5) will hold for all

$$a \in \mathrm{Log}(r(I)p^{l(n,\lambda)})\setminus \mathcal{N}_{l(n,\lambda)}(\partial W(p), \frac{\epsilon_{k-1}}{3}l(n,\lambda)),$$

provided n is large enough; the constant $\bar{K} = \bar{K}_k$ is imposed by (9.5), but the constant $\theta = \theta_k$ in (9.4) may be made arbitrarily small. To see that (9.7) may

be applied to $P(n, \lambda, i)$, suppose $W_{l(n,\lambda)}(I, J, w)$ is contained in $W_{l(n,\lambda)}(k-1, \lambda - 1, \ldots, \lambda - \epsilon(k-1)n - 1)$ but not in
$$W_{l(n,\lambda)}(k-1, \lambda - 1 - \bar{D}, \ldots, \lambda - \epsilon(k-1)n - 1 - \bar{D}).$$
Recall that $P(n, \lambda, i) = P(n, \lambda+1, i, 0) = P(n, \lambda+1, i, d_{k-1} - 1)$ to consider the decomposition (∗∗) of $W_{l(n,\lambda)}(I, J, w) \cap P(n, \lambda, i)$. It follows from (9.5) and (9.8) that the total cardinality of the terms of (∗∗) corresponding to $h \geq h_1$ may be made an arbitrarily small fraction of $|W_{l(n,\lambda)}(I, J, w)|$ by picking large h_1. For each $W(\lambda - h, m, J_0, w_0, \eta, i)$ with $h < h_1$, we have $F \in \mathcal{F}_m(p)$ such that
$$\frac{\nu(\lambda-h,m,J_0,w_0,\eta,i)}{\pi(l(n,\lambda),w+a_J,i)} = \frac{\nu(\lambda-h,m,J_0,w_0,\eta,i)}{\pi(l(n,\lambda-h,d_m-1),w_0+a_{F,J_0},i)} \frac{\pi(l(n,\lambda,d_m-1),w_0+a_{F,J_0},i)}{\pi(l(n,\lambda),w+a_J,i)},$$
and (9.9)(ii) and (9.4) show that this quantity will be close to 1 provided n is large. It follows that, for large n, (9.7) will apply to $P(n, \epsilon(k-1)n + \frac{2}{3}\epsilon_{k-1}n, i)$ and complete the proof of (3.2). It remains for us to prove (9.9).

PROOF OF (9.9). Let $j \in \{0, 1, \ldots, k-1\}$. Assume that (9.9) has been established for $j - 1$. Then we have $\epsilon_m, K_m, \bar{K}_m, \theta_m, 0 \leq m \leq j - 1$, such that for any $\xi > 0$ statements (i) and (ii) are valid in the case of $j - 1$ as long as n is large. We explain how to choose $\epsilon_j, K_j, \bar{K}_j, \theta_j$ so that (i) and (ii) hold for j.

Pick $\epsilon_j, \bar{K}_j$ so that (9.5) holds with $\bar{K} = \bar{K}_j$ on $\mathcal{N}_n(F, (k-j)\epsilon_j n) \setminus \mathcal{N}_n(\partial F, \frac{\epsilon_{j-1}}{3}n)$ for all $F \in \mathcal{F}_j(p)$ and large n. Fixing $\bar{K}_j$, we will decrease ϵ_j in order for (9.1) and (9.4) to hold for suitable K_j, θ_j. Since $P^{(j)}(n, \lambda, i, \mu) = P^{(j-1)}(n, \lambda, i, \mu)$ for $\lambda \leq \epsilon(j)n$, we consider $\lambda \geq \epsilon(j)n + 1$. To proceed by induction, assume (i) and (ii) are valid for lower values of λ. For $F \in \mathcal{F}_j(p)$ and $a' \in \mathcal{N}_{l(n,\lambda,d_j-1)}(F, \lambda - \epsilon(j)n - 1) \setminus \mathcal{N}_1^{(j)}(n, (\lambda-1)\tilde{d} + d_j - 1)$, let $W_{l(n,\lambda,d_j-1)}(I, J, w)$ and $\eta_0 \in W_L(J)$ be such that we need paths from $W_{l(n,\lambda,d_j-1)}(I, J, w)\eta_0$ in transferring $x^{a'} p_F^L$. In particular, $W_{l(n,\lambda,d_j-1)}(I, J, w) \subset W_{l(n,\lambda,d_j-1)}(F, \lambda-\epsilon(j)n-1)$, J lies in a principal F-component $\mathcal{C}$ for which (9.5) is valid, and η_0 is an A_F-path. If
$$(w + \mathrm{Log}(r(J))) \cap \mathcal{N}_1^{(j)}(n, (\lambda-1)\tilde{d} + d_j - 1) = \emptyset$$
then, by construction, $\widetilde{P}^{(j)}(n, \lambda, i, d_j)$ contains at least
$$\lfloor \pi(l(n, \lambda, d_j - 1), w + a_{F,J}, i) |W_{l(n,\lambda,d_j-1)}(I, J, w)| \rfloor$$
elements of $W_{l(n,\lambda,d_j-1)}(I, J, w)\eta_0$. For $j \geq 1$, it is quite possible that $a_{F,J}$ is not the only element of $\mathrm{Log}(r_F(J))$ but, using (9.4) to pick suitable ϵ_j and θ_j, we can ensure that
$$\pi(l(n, \lambda, d_j - 1), w + a_{F,J}, i)/\pi(l(n, \lambda, d_j - 1), w + b, i)$$
is as close to 1 as we wish for all $b \in \mathrm{Log}(r_F(J))$. If the above intersection is nonempty, that is, if
$$W_{l(n,\lambda,d_j-1)}(I, J, w) \subset W_{l(n,\lambda,d_j-1)}(j, \lambda - 2, \lambda - \epsilon(1)n - 2, \ldots, \lambda - \epsilon(j)n - 2),$$
then we consider the decomposition (∗∗) of $W_{l(n,\lambda,d_j-1)}(I, J, w) \cap P^{(j)}(n, \lambda, i, d_j - 1)$. Since J lies in a principal component $\mathcal{C}$ for which (9.5) is valid, (9.8) implies that the total cardinality of the terms of (∗∗) corresponding to $h \geq h_1$ may be made an

arbitrarily small fraction of $|W_{l(n,\lambda,d_j-1)}(I,J,w)|$ by picking h_1 to be sufficiently large. The terms of (**) with $h < h_1$ are given by

$$W^{(j)}(\lambda - h, m, J_0, w_0, \eta, i)$$
$$= \{\gamma \in W_{l(n,\lambda-h,d_m-1)}(I, J_0, w_0) : \gamma\eta \in P^{(j)}(n, \lambda, i, d_j - 1)\},$$

where $W_{l(n,\lambda-h,d_m-1)}(I, J, w_0)$ does not intersect

$$W_{l(n,\lambda-h,d_m-1)}(j, \lambda - h - 2, \ldots, \lambda - \epsilon(j)n - h - 2)$$

but is contained in $W_{l(n,\lambda-h,d_m-1)}(G, \lambda - h - \epsilon(m)n - 1)$ for some m-dimensional face G of F. There are two possibilities. One is that $w_0 \cdot v_F + \delta_F(J_0) < \delta_F(I) + l(n, \lambda - h, d_m - 1)\delta_F + \lambda - \epsilon(j)n - 1$ (which is equivalent to η not being an A_F-path): Using (1) and (9.1) and picking $K = K_j$ to be sufficiently large, the total cardinality of these sets may be made an arbitrarily small fraction of $|W_{l(n,\lambda,d_j-1)}(I, J, w)|$. The second possibility is that $w_0 \cdot v_F + \delta_F(J_0) = \delta_F(I) + l(n, \lambda - h, d_m - 1)\delta_F + \lambda - \epsilon(j)n - 1$ (which is equivalent to η being an A_F-path): For these sets (9.4) and, using our inductive assumption, (9.9)(ii) imply that the ratio

$$\nu^{(j)}(\lambda - h, m, J_0, w_0, \eta, i)/\pi(l(n, \lambda, i, d_j - 1), w + a_{F,J}, i)$$

is close to 1. Putting these facts together, we can be sure that

$$\frac{|W_{l(n,\lambda,d_j-1)}(I, J, w) \cap P^{(j)}(n, \lambda, i, d_j - 1)|}{\pi(l(n, \lambda, d_j - 1), a', i)|W_{l(n,\lambda,d_j-1)}(I, J, w)|}$$

is close to 1. Moreover, for large K_j and small θ_j, the number of times we need to transfer $x^{a'}p_F^L$ while making corrections to $\widetilde{P}^{(j)}(n, \lambda, i, d_j)$ will be an arbitrarily small fraction of $|W_{l(n,\lambda,d_j-1)}(I, J, w)|$. This proves (i) for j and, also, (ii) for j and $m \in \{0, 1, \ldots, j-1\}$. In addition, in the case $m = j$, we can ensure that

$$|\nu^{(j)}(\lambda, j, J, w, \eta, i)/\pi(l(n, \lambda, d_j - 1), w + a_{F,J}, i) - 1| < \frac{1}{8}$$

for all $\epsilon(j)n < \lambda \le \epsilon(j)n + (k-j)\epsilon_j n$. From this inequality and (9.8) we obtain (9.9)(ii) for $m = j$ by arguing as in the proof of (7.11)(i). For $F \in \mathcal{F}_j(p)$, to cover all of

(†) $\quad W_{l(n,\lambda,d_j-1)}(F, \lambda - \epsilon(j)n - 1)\setminus W_{l(n,\lambda,d_j-1)}(j, \lambda - 2, \ldots, \lambda - \epsilon(j)n - 2),$

go through the argument $(k-j)$ times: For $m = j+1, \ldots, k$ pick suitable $\epsilon'_m > 0$ and, for $G \in \mathcal{F}_m(p)$, apply the argument of (7.11)(i) to the intersection of (†) with

$$W_{l(n,\lambda,d_j-1)}(G, \epsilon'_m n)\setminus W_{l(n,\lambda,d_j-1)}(\partial G, \frac{\epsilon'_{m-1}}{3}n),$$

making use of (9.1) and (9.4) on $\mathcal{N}_{l(n,\lambda,d_j-1)}(G, \epsilon'_m n)\setminus \mathcal{N}_{l(n,\lambda,d_j-1)}(\partial G, \frac{\epsilon'_{m-1}}{3}n)$. (Start by taking $\epsilon'_j = \epsilon_j/3$.) $\quad\square$

Bibliography

[A1] J. Ashley, Bounded-to-1 factors of an aperiodic shift of finite type are 1-to-1 almost everywhere factors also, *Ergod. Th. and Dynam. Sys.* **10** (1990), 615–626.

[A2] J. Ashley, Resolving factor maps for shifts of finite type with equal entropy, *Ergod. Th. and Dynam. Sys.* **11** (1991), 219–240.

[AMT] J. Ashley, B. Marcus and S. Tuncel, The classification of one-sided Markov chains, *Ergod. Th. and Dynam. Sys.* **17** (1997), 269–295.

[BMT] M. Boyle, B. Marcus and P. Trow, Resolving maps and the dimension group for shifts of finite type, *Mem. Amer. Math. Soc.* **377** (1987).

[BT] M. Boyle and S. Tuncel, Regular isomorphism of Markov chains is almost topological, *Ergod. Th. and Dynam. Sys.* **10** (1990), 89–100.

[FP] R. Fellgett and W. Parry, Endomorphisms of a Lebesgue space II, *Bull. London Math. Soc.* **7** (1975), 151–158.

[FO] N. Friedman and D. Ornstein, On the isomorphism of weak Bernoulli transformations, *Adv. in Math.* **5** (1970), 365–394.

[H1] D. Handelman, Positive polynomials and product type actions of compact groups, *Mem. Amer. Math. Soc.* **320** (1985).

[JKKMS] A. del Junco, M. Keane, B. Kitchens, B. Marcus and L. Swanson, Continuous homomorphisms of Bernoulli schemes, *Progress in Math.*, Vol. 10, Birkhäuser, Boston, Basel and Stuttgart, 1981, pp. 91–111.

[K] B. Kitchens, Ph.D. thesis, University of North Carolina, Chapel Hill, 1981.

[K1] W. Krieger, On the subsystems of topological Markov chains, *Ergod. Th. and Dynam. Sys.* **2** (1982), 195–202.

[K2] W. Krieger, On the finitary isomorphisms of Markov shifts that have finite expected coding time, *Z. Wahrscheinlichkeitstheorie*, **65** (1983), 323–328.

[KMT] W. Krieger, B. Marcus and S. Tuncel, On automorphisms of Markov chains, *Trans. Amer. Math. Soc.* **333** (1992), 531–565.

[LM] D. Lind and B. Marcus, *An Introduction to Symbolic Dynamics and Coding*, Cambridge Univ. Press, Cambridge, 1995.

[M] B. Marcus, Factors and extensions of full shifts, *Monatshefte Math.* **88** (1979), 239–247.

[MT1] B. Marcus and S. Tuncel, The weight-per-symbol polytope and scaffolds of invariants associated with Markov chains, *Ergod. Th. and Dynam. Sys.* **11** (1991), 129–180.

[MT2] B. Marcus and S. Tuncel, Entropy at a weight-per-symbol and embeddings of Markov chains, *Invent. Math.* **102** (1990), 235–266.

[MT3] B. Marcus and S. Tuncel, Matrices of polynomials, positivity, and finite equivalence of Markov chains, *J. Amer. Math. Soc.* **6** (1993), 131–147.

[MT4] B. Marcus and S. Tuncel, On large powers of positive polynomials in several variables, in this issue of *Mem. Amer. Math. Soc.*

[O] D. Ornstein, *Ergodic Theory, Randomness and Dynamical Systems*, Yale Univ. Press, New Haven, 1974.

[PS] W. Parry and K. Schmidt, Natural coefficients and invariants for Markov shifts, *Invent. Math.* **76** (1984), 15–32.

[PT1] W. Parry and S. Tuncel, On the classification of Markov chains by finite equivalence, *Ergod. Th. and Dynam. Sys.* **1** (1981), 303–335.

[PT2] W. Parry and S. Tuncel, *On the stochastic and topological structure of Markov chains*, Bull. London Math. Soc. **14** (1982), 16–27.

[PT3] W. Parry and S. Tuncel, *Classification Problems in Ergodic Theory*, Cambridge Univ. Press, Cambridge, 1982.

[S] E. Seneta, *Non-negative Matrices and Markov Chains*, Springer, New York, 1980.

[T] P. Trow, *Resolving maps which commute with a power of the shift*, Ergod. Th. and Dynam. Sys. **6** (1986), 281–293.

[T1] S. Tuncel, *Conditional pressure and coding*, Israel J. Math. **39** (1981), 101–112.

[T2] S. Tuncel, *A dimension, dimension modules and Markov chains*, Proc. London Math. Soc. **46** (1983), 100–116.

[T3] S. Tuncel, *Faces of Markov chains and matrices of polynomials*, Contemp. Math., Vol. 135, Amer. Math. Soc., Providence, 1992, pp. 391–422.

[T4] S. Tuncel, *Coefficient rings for beta function classes of Markov chains*, to appear in Ergod. Th. and Dynam. Sys..

[W] R. Williams, *Classification of subshifts of finite type*, Ann. Math. **98** (1973), 120–153; errata, Ann. Math. **99** (1974), 380–381.

Part B

On Large Powers of Positive Polynomials in Several Variables

Introduction

Let p be a Laurent polynomial in the variables $x_1, \ldots, x_k$. For $w = (w_1, \ldots, w_k)$ in $\mathbb{Z}^k$, write $x^w = x_1^{w_1} \cdots x_k^{w_k}$ and denote the coefficient of x^w in p by p_w. Then
$$p = \sum_{w \in \mathbb{Z}^k} p_w x^w,$$
and p_w are nonzero for only finitely many $w \in \mathbb{Z}^k$. Let
$$\mathrm{Log}(p) = \{w \in \mathbb{Z}^k : p_w \neq 0\}.$$
The *Newton polyhedron* $W(p)$ of p is given by the convex hull of $\mathrm{Log}(p)$. Denote the collection of nontrivial faces of $W(p)$ by $\mathcal{F}(p)$. For $F \in \mathcal{F}(p)$, let
$$p_F = \sum_{w \in F \cap \mathbb{Z}^k} p_w x^w$$
and call p_F the *F-face* of p. Let $\Delta(p)$ be the subgroup of $\mathbb{Z}^k$ generated by $\{a - b : a, b \in \mathrm{Log}(p)\}$, and pick any $c_p \in \mathrm{Log}(p)$ to distinguish a coset $c_p + \Delta(p)$ of $\Delta(p)$.

It is easy to see that $\mathrm{Log}(p^n) \subset nW(p) \cap (nc_p + \Delta(p))$. One of our results will determine the set $\mathrm{Log}(p^n)$ for large powers p^n in the case p has nonnegative coefficients. If $H \geq 0$ and $S \subset \mathbb{R}^k$, we use the usual norm and distance in $\mathbb{R}^k$ to write $B(S, H) = \{w \in \mathbb{R}^k : \mathrm{dist}(w, S) < H\}$. For $F \in \mathcal{F}(p)$ we denote by ∂F the boundary of F taken in its affine hull. Equivalently, ∂F is the union of all $G \in \mathcal{F}(p)$ which are proper subsets of F. We put
$$\mathrm{Int}(p^n, H) = (W(p^n) \setminus B(\partial W(p^n), H)) \cap (nc_p + \Delta(p)).$$
In other words, $\mathrm{Int}(p^n, H)$ is the set of elements of $nc_p + \Delta(p)$ remaining in $W(p^n) = nW(p)$ after we remove a band of width H from around the boundary. For $p \in \mathbb{R}^+[x_1^\pm, \ldots, x_k^\pm]$, we will first show that there exists $H(p) \geq 0$ such that $\mathrm{Int}(p^n, H(p)) \subset \mathrm{Log}(p^n)$ for all $n \in \mathbb{N}$. We will then prove the following.

THEOREM 1.1. *Let $p \in \mathbb{R}^+[x_1^\pm, \ldots, x_k^\pm]$. Suppose that, for each $F \in \mathcal{F}(p)$, we have constants $K_F \geq H(p_F)$ and $D_F \geq 0$. Then there exist $H_F \geq K_F$, $N \in \mathbb{N}$ and finite subsets $T(F) \subset \mathbb{Z}^k$ such that $T(W(p)) = \{0\}$ and*
$$\mathrm{Log}(p^n) = \bigcup_{F \in \mathcal{F}(p)} (\mathrm{Int}(p_F^n, H_F + D_F) + T(F)) = \bigcup_{F \in \mathcal{F}(p)} (\mathrm{Int}(p_F^n, H_F) + T(F))$$
for all $n \geq N$.

In particular, other than magnification of the materials used, for large n the sets $\mathrm{Log}(p^n)$ are built in the same way: For each $F \in \mathcal{F}(p)$ cut from the lattice $\Delta(p_F)$ (or its copy $nc_{p_F} + \Delta(p_F)$) exactly $|T(F)|$ pieces of the right size in the shape of $W(p_F)$, translate them according to $T(F)$, glue them together, and you have $\text{Log}(p^n)$. (The right size is obtained by magnifying $W(p_F)$ by n and erasing everything within H_F of the boundary.)

Theorem (1.1) will be proved in section 2. In section 5 we will prove the following three estimates on the relative sizes of coefficients of p^n for large n and $p \in \mathbb{Z}^+[x_1^\pm, \dots, x_k^\pm]$. For $F \in \mathcal{F}(p)$, we write $c_F = c_{p_F}$.

PROPOSITION 1.2. *Let $p \in \mathbb{Z}^+[x_1^\pm, \dots, x_k^\pm]$ and let F be a proper face of $W(p)$. Let $L \in \mathbb{N}$ and $K, \bar{\epsilon}_F > 0$. There exist $\epsilon_F > 0$ and $N \in \mathbb{N}$ such that for $a \in \mathbb{Z}^k$ the inequality*

$$(p^n)_{a+b} \geq K(p^n)_a$$

holds whenever $n \geq N$, $\frac{a}{n} \in B(F, \epsilon_F) \backslash B(\partial F, \bar{\epsilon}_F)$ and $b \in (\text{Log}(p^L) \backslash \text{Log}(p_F^L)) - \text{Log}(p_F^L)$.

PROPOSITION 1.3. *Let $p \in \mathbb{Z}^+[x_1^\pm, \dots, x_k^\pm]$ and $F \in \mathcal{F}(p)$. Let $\bar{\epsilon}_F, \theta > 0$ and $D \in \mathbb{N}$. There exist $\epsilon_F > 0$ and $N \in \mathbb{N}$ such that*

$$\left| \frac{(p^n)_a / (p^n)_{a+b}}{(p^n)_{a+u} / (p^n)_{a+u+b}} - 1 \right| < \theta$$

whenever $n \geq N$ and $a, b, u \in \mathbb{Z}^k$ are such that $a, a+u \in \text{Log}(p^n)$, $b \in \Delta(p_F)$, $\frac{a}{n} \in B(F, \epsilon_F) \backslash B(\partial F, \bar{\epsilon}_F)$ and $\|u\| \leq D$.

PROPOSITION 1.4. *Let $p \in \mathbb{Z}^+[x^\pm, \dots, x_k^\pm]$ and $F \in \mathcal{F}(p)$. Let $\epsilon_F > 0$, $l \in \mathbb{Z}$ and $D \in \mathbb{N}$. There exist $\epsilon_F > 0$ and $K, N \in \mathbb{N}$ such that*

$$\frac{1}{K}(p^{n+l})_{a+lc_F+b} \leq (p^n)_a \leq K(p^{n+l})_{a+lc_F+b}$$

whenever $n \geq N$ and $a, b \in \mathbb{Z}^k$ are such that $\frac{a}{n} \in B(F, \epsilon_F) \backslash B(\partial F, \bar{\epsilon}_F)$, $b \in \Delta(p_F)$ and $\|b\| \leq D$.

(1.1)–(1.4) are used in [**MT4**] in the context that lead us to them, that of coding problems in ergodic theory. The presentation given in the present paper is independent of coding and ergodic theory, and should be accessible to anyone with an interest in powers of polynomials. The estimates (1.2)–(1.4) are not definitive. They are what we needed for the work in [**MT4**]. The method of proof and the more technical results we establish, in particular theorems (3.1) and (4.7), may be objects of greater interest. They enable estimates to be carried out near any $a \in \text{Log}(p^n)$, and we suspect they can be used to obtain further results on coefficients. We consider $\frac{a}{n} \in W(p)$ and cover $W(p)$ by using induction on the affine dimension of $F \in \mathcal{F}(p)$ to pick sets which consist of (small) neighbourhoods of F from which previously chosen neighbourhoods of ∂F have been removed. We pick ϵ_F small enough for $B(F, \epsilon_F) \backslash B(\partial F, \bar{\epsilon}_F)$ to inherit from p_F properties that hold on compact subsets of the relative interior of $W(p_F)$. We rely on the notion of entropy at a weight-per-symbol and the related equilibrium distributions, which were introduced in [**MT2**]. Our setting is simpler than that of [**MT2**] (we are dealing with the Bernoulli case, as opposed to Markov), and allows more general statements. In particular, we show in theorem (3.1) that the mapping $w \mapsto s(w)$ of $w \in W(p)$ to

1. INTRODUCTION

the corresponding equilibrium distribution $s(w)$ is C^1 on $W(p)$. This result is close to the edge of what is possible: Its analogue is not valid in the (Markov) setting of [**MT2**]. In the present (Bernoulli) setting, examples show that $w \mapsto s(w)$ need not be C^2, and that closely related maps (in particular the intermediate map $w \mapsto x(w)$ of section 3) need not be differentiable. Equilibrium distributions are developed in section 3 and in section 4, culminating in their relation in (4.7) to the coefficients of p^n. The proofs of (1.2)–(1.4) are then given in section 5.

While the idea of reference to p_F is common to (1.1)–(1.4), the reader should note that the neighbourhoods used in (1.1) differ from those of (1.2)–(1.4). In (1.1) a fixed distance H is used for all n. In (1.2)–(1.4) we pick neighbourhoods in $W(p)$ and blow them up by a factor of n to obtain neighbourhoods in $W(p^n)$.

As far as we are aware, F-faces p_F were introduced by Handelman [**H1**]. They were further utilized by Handelman in [**H2**] and other papers, and by the authors in [**MT3**]. Extensions to matrices of nonnegative polynomials were carried out and exploited in [**MT1, T, MT4**].

Structure of Log(p^n)

In this section, after establishing the constants $H(p_F)$ used in the statement of (1.1), we will prove (1.1).

PROPOSITION 2.1. *Let $p \in \mathbb{R}^+[x_1^{\pm}, \ldots, x_k^{\pm}]$.*
(i) $\mathrm{Log}(p^n) \subset nW(p) \cap (nc_p + \Delta(p))$.
(ii) *There exists $H(p) \geq 0$ such that*
$$\mathrm{Int}(p^n, H(p)) \subset \mathrm{Log}(p^n)$$
for all $n \in \mathbb{N}$.

PROOF. (i) That $\mathrm{Log}(p^n) \subset nW(p) = W(p^n)$ is clear. The inclusion $\mathrm{Log}(p^n) \subset nc_p + \Delta(p)$ follows from the fact that $\mathrm{Log}(x^{-c_p}p) \subset \Delta(x^{-c_p}p) = \Delta(p)$.
(ii) Dividing p by a suitable monomial and making a change of variables, we may assume that $0 \in \mathrm{Log}(p) \subset \Delta(p) = \mathbb{Z}^k$ and that 0 is an extreme point of $W(p)$. Letting $f(j) \in \mathbb{R}^k$ denote the j-th canonical basis vector, find $m_{j,w} \in \mathbb{Z}$ such that

(2.1) $$f(j) = \sum_{w \in \mathrm{Log}(p)} m_{j,w} w$$

and put
$$M = \max\{|m_{j,w}| : 1 \leq j \leq k, w \in \mathrm{Log}(p)\}.$$

List the extreme points of $W(p)$ as $w(0) = 0, w(1), \ldots, w(d)$. Let $e(0) = 0 \in \mathbb{R}^d$ and, for $1 \leq i \leq d$, let $e(i)$ be the i-th canonical basis vector in $\mathbb{R}^d$. The convex hull of $e(0), e(1), \ldots, e(d)$ is the d-dimensional simplex in $\mathbb{R}^d$ which we denote by S. Define a linear map $\Phi : \mathbb{R}^d \to \mathbb{R}^k$ by letting $\Phi(e(i)) = w(i)$ for $i \in \{0, 1, \ldots, d\}$. Since

$$\mathrm{Log}(p^n) = \{\sum_{w \in \mathrm{Log}(p)} n_w w : n_w \in \mathbb{Z}^+, \sum_{w \in \mathrm{Log}(p)} n_w = n\},$$

we have $\Phi(nS) = W(p^n)$ and $\Phi(nS \cap \mathbb{Z}^d) \subset \mathrm{Log}(p^n)$. It follows from $\Delta(p) = \mathbb{Z}^k$ that $\Phi(\mathbb{Z}^d)$ is a subgroup of finite index in $\mathbb{Z}^k$. Letting

$$K = k \, |\mathbb{Z}^k / \Phi(\mathbb{Z}^d)|,$$

observe that for any $u \in \mathbb{Z}^k$ we can find $v \in \Phi(\mathbb{Z}^d)$ such that, in the 1-norm, $\|u - v\|_1 \leq K$. Put $l = 2MK|Log(p)|$ and

$$W_{n-l} = \Phi((n-l)S) + \sum_{w \in \mathrm{Log}(p)} MKw.$$

Since $0 \in \text{Log}(p)$, the polytope W_{n-l} is contained in $W(p^n)$. Moreover, W_{n-l} covers all of $W(p^n)$ except $B(\partial W(p), H)$, where H is independent of n. We complete the proof of (ii) by verifying that any $u \in \mathbb{Z}^k$ which lies, in the 1-norm, within K of an element of

$$\Phi\left((n-l)S \cap \mathbb{Z}^d\right) + \sum_{w \in \text{Log}(p)} MKw$$

must belong to $\text{Log}(p^n)$: Let $v \in \Phi((n-l)S \cap \mathbb{Z}^d)$ and $l_j \in \mathbb{Z}$ be such that $\sum_{j=1}^k |l_j| \leq K$ and

$$u = v + \sum_{w \in \text{Log}(p)} MKw + \sum_{j=1}^k l_j f(j).$$

Then, using (2.1),

$$u = v + \sum_{w \in \text{Log}(p)} \left(MK + \sum_{j=1}^k l_j m_{j,w}\right) w.$$

Note that in the last sum the coefficient of each $w \in \text{Log}(p)$ is nonnegative as a result of our choice of M. In addition the sum of these coefficients is bounded above by $l = 2MK|\text{Log}(p)|$. Since $v \in \text{Log}(p^{n-l})$ and $0 \in \text{Log}(p)$, it follows that $u \in \text{Log}(p^n)$. $\square$

In the case of one variable $y = x_1$, theorem (1.1) is not new and follows immediately from (2.1):

COROLLARY 2.2. *For $p \in \mathbb{R}^+[y^\pm]$, let $m = \min(\text{Log}(p))$, $M = \max(\text{Log}(p))$ and $H = H(p)$. There exist subsets $T_0 \subset \{0, \ldots, H\}$ and $T_\infty \subset \{-H, \ldots, 0\}$ such that*

$$\text{Log}(p^n) = (T_0 + nm) \cup \left([nm + H, nM - H] \cap (nm + \Delta(p))\right) \cup (T_\infty + nM)$$

for all $n \geq H = H(p)$.

PROOF. Apply (2.1) with $k = 1$ and take

$$T_0 = [0, H] \cap \{a_1 + \cdots + a_H - Hm : a_1, \ldots, a_H \in \text{Log}(p)\}$$

and

$$T_\infty = [-H, 0] \cap \{a_1 + \cdots + a_H - HM : a_1, \ldots, a_H \in \text{Log}(p)\}. \quad \square$$

We denote the affine dimension of $F \in \mathcal{F}(p)$ by $\dim(F)$. Recall that $F \in \mathcal{F}(p)$ is called a *facet* of $W(p)$ if it has codimension 1, that is, if $\dim(W(p)) - \dim(F) = 1$. Let us denote the collection of facets of $W(p)$ by $\widetilde{\mathcal{F}}(p)$.

LEMMA 2.3. *For $p \in \mathbb{R}^+[x_1^\pm, \ldots, x_k^\pm]$ and $K \geq H(p)$, there exists $L = L(p, K)$ with the following property. For $l \geq L$ and each facet F of $W(p)$ we can find a finite set $T(F) = T(F, K, l)$ such that*

$$\text{Log}(p^n) = \text{Int}(p^n, K) \cup \bigcup_{F \in \widetilde{\mathcal{F}}(p)} (\text{Log}(p_F^{n-l}) + T(F))$$

for all large n.

PROOF. For $F \in \widetilde{\mathcal{F}}(p)$ find $v_F \in \mathbb{Z}^k$ which exposes F in the sense that

$$a_F \equiv \min\{W(p) \cdot v_F\} = F \cdot v_F < \min\{(\mathrm{Log}(p)\setminus F) \cdot v_F\}.$$

Put

$$b_F = \min\{(\mathrm{Log}(p)\setminus F) \cdot v_F\} - a_F > 0.$$

By (2.1) we have $\mathrm{Log}(p^n)\setminus B(\partial W(p^n), K) = \mathrm{Int}(p^n, K)$. To capture the elements of $\mathrm{Log}(p^n)$ that lie in $B(\partial W(p^n), K)$, find a constant b such that the intersection

$$I(p^n) \equiv \bigcap_{F \in \widetilde{\mathcal{F}}(p)} \{w \in nc_p + \Delta(p) : w \cdot v_F \geq na_F + b\}$$

is contained in $\mathrm{Int}(p^n, K)$. Let

$$L = 1 + \max\{\lfloor b/b_F \rfloor : F \in \widetilde{\mathcal{F}}(p)\}.$$

Consider $l \geq L$. Recall that

$$\mathrm{Log}(p^n) = \{\sum_{w \in \mathrm{Log}(p)} n_w w : n_w \in \mathbb{Z}^+, \sum_{w \in \mathrm{Log}(p)} n_w = n\}.$$

A sum w of n elements of $\mathrm{Log}(p)$ has $w \cdot v_F \geq na_F$, with equality if and only if all of the n elements come from $\mathrm{Log}(p) \cap F$. Each time we use an element of $\mathrm{Log}(p)\setminus F$, the inner product with v_F increases by at least b_F. It follows that the sets

$$T(F) = \{w_1 + \cdots + w_l : w_1, \ldots, w_l \in \mathrm{Log}(p)\}$$

are such that

$$\mathrm{Int}(p^n, K) \cup \bigcup_{F \in \widetilde{\mathcal{F}}(p)} (\mathrm{Log}(p_F^{n-l}) + T(F))$$

$$= I(p^n) \cup \bigcup_{F \in \widetilde{\mathcal{F}}(p)} (\mathrm{Log}(p_F^{n-l}) + T(F)) = \mathrm{Log}(p^n). \quad \square$$

PROOF OF (1.1). Apply (2.3) to p and $K = K_{W(p)} + D_{W(p)}$ to find $l \in \mathbb{N}$ and, for each $F \in \widetilde{\mathcal{F}}(p)$, a finite set $T(F)$ with

(2.2)
$$\begin{aligned}\mathrm{Log}(p^n) &= \mathrm{Int}(p^n, K_{W(p)} + D_{W(p)}) \cup \bigcup_{F \in \widetilde{\mathcal{F}}(p)} (\mathrm{Log}(p_F^{n-l}) + T(F)) \\ &= \mathrm{Int}(p^n, K_{W(p)}) \cup \bigcup_{F \in \widetilde{\mathcal{F}}(p)} (\mathrm{Log}(p_F^{n-l}) + T(F))\end{aligned}$$

for all large n. We will use induction on the codimension. Writing $d = \dim(W(p))$, suppose $j \in \{1, \ldots, d-1\}$ is such that the following holds: For each $F \in \mathcal{F}(p)$ with

$\dim(F) \geq j$ we have a finite set $T(F) \subset \mathbb{Z}^k$, for each $F \in \mathcal{F}(p)$ with $\dim(F) \geq j+1$ we have $H_F \geq K_F$, and we have $l \in \mathbb{N}$ such that

(2.3)

$\mathrm{Log}(p^n)$

$$= \bigcup_{\dim(F) \geq j+1} (\mathrm{Int}(p_F^n, H_F + D_F) + T(F)) \cup \bigcup_{\dim(F)=j} (\mathrm{Log}(p_F^{n-l}) + T(F))$$

$$= \bigcup_{\dim(F) \geq j+1} (\mathrm{Int}(p_F^n, H_F) + T(F)) \cup \bigcup_{\dim(F)=j} (\mathrm{Log}(p_F^{n-l}) + T(F))$$

for all large n. We will show that a similar statement then holds for $j - 1$. With $H_{W(p)} = K_{W(p)}$ and $T(W(p)) = \{0\}$, equation (2.2) starts the induction for $j = d - 1$. For the inductive step pick, for each $F \in \mathcal{F}(p)$ with $\dim(F) = j$, constants H_F, H'_F such that $H'_F \geq H_F + D_F \geq H_F \geq K_F$ and

(2.4) $\quad \mathrm{Int}(p_F^{n-l}, H'_F) \subset \mathrm{Int}(p_F^n, H_F + D_F) \subset \mathrm{Int}(p_F^n, H_F) \subset \mathrm{Int}(p_F^{n-l}, H(p_F))$.

Applying (2.3) to p_F and $K = H'_F$ and letting $l' = \max\{L(p_F, H'_F) : \dim(F) = j\}$ we have for each $F \in \mathcal{F}(p)$ with $\dim(F) = j$ and each $G \in \widetilde{\mathcal{F}}(p_F)$ a finite set $T(F, G)$ such that, for all large n,

$\mathrm{Log}(p_F^{n-l})$

$$= \mathrm{Int}(p_F^{n-l}, H'_F) \cup \bigcup_{G \in \widetilde{\mathcal{F}}(p_F)} (\mathrm{Log}(p_G^{n-l-l'}) + T(F, G))$$

$$= \mathrm{Int}(p_F^n, H_F + D_F) \cup \bigcup_{G \in \widetilde{\mathcal{F}}(p_F)} (\mathrm{Log}(p_G^{n-l-l'}) + T(F, G)) \qquad \text{(by (2.4))}$$

$$= \mathrm{Int}(p_F^n, H_F) \cup \bigcup_{G \in \widetilde{\mathcal{F}}(p_F)} (\mathrm{Log}(p_G^{n-l-l'}) + T(F, G)).$$

For each $G \in \mathcal{F}(p)$ with $\dim(G) = j - 1$, let

$$T(G) = \bigcup_{\substack{F \in \mathcal{F}(p), \dim(F)=j, \\ G \subset F}} (T(F, G) + T(F)).$$

Then, substituting for $\mathrm{Log}(p_F^{n-l})$ in equation (2.3), we have

$\mathrm{Log}(p^n)$

$$= \bigcup_{\dim(F) \geq j} (\mathrm{Int}(p_F^n, H_F + D_F) + T(F)) \cup \bigcup_{\dim(G)=j-1} (\mathrm{Log}(p_G^{n-l-l'}) + T(G))$$

$$= \bigcup_{\dim(F) \geq j} (\mathrm{Int}(p_F^n, H_F) + T(F)) \cup \bigcup_{\dim(G)=j-1} (\mathrm{Log}(p_G^{n-l-l'}) + T(G)).$$

This provides the inductive step. Denote the set of extreme points of $W(p)$ by $\mathrm{Ext}(W(p))$. The final application of the inductive step is to $j = 1$. It yields $l \in \mathbb{N}$, for each $F \in \mathcal{F}(p)$ a finite set $T(F) \subset \mathbb{Z}^k$, and for each $F \in \mathcal{F}(p)$ with $\dim(F) = 1$ a constant $H_F \geq K_F$ such that, for all large n,

$$\mathrm{Log}(p^n)$$
$$= \bigcup_{\dim(F) \geq 1} (\mathrm{Int}(p_F^n, H_F + D_F) + T(F)) \cup \bigcup_{w \in \mathrm{Ext}(W(p))} ((n-l)w + T(w))$$

$$= \bigcup_{\dim(F) \geq 1} (\mathrm{Int}(p_F^n, H_F) + T(F)) \cup \bigcup_{w \in \mathrm{Ext}(W(p))} ((n-l)w + T(w)).$$

Note that when $F = \{w\}$ consists of an extreme point, $\mathrm{Int}(p_F^n, H) = nw$ for all $H \geq 0$. Since $(n-l)w + T(w) = nw + (T(w) - lw)$, the theorem is proved. $\square$

Entropy and equilibrium distributions for $\mathbf{w} \in \mathbf{W(p)}$

In this section we consider entropy at each $w \in W(p)$ and establish the existence and continuous differentiability of equilibrium distributions, or states, $s(w)$ for $w \in W(p)$. In section 4 we will relate the distributions $s(w)$ to the coefficients of p^n and, in section 5, use this relationship to prove (1.2)–(1.4). For the remainder of the paper, we fix $p \in \mathbb{Z}^+[x_1^\pm, \ldots, x_k^\pm]$ and assume without loss of generality that $\mathrm{Log}(p) \subset \Delta(p) = \mathbb{Z}^k$.

Putting $d = p(1, \ldots, 1)$, let $a(1), \ldots, a(d) \in \mathbb{Z}^k$ be such that

$$p = \sum_{i=1}^{d} x^{a(i)}.$$

Let S denote the standard $(d-1)$-simplex in $\mathbb{R}^d$, and $\langle S \rangle$ the hyperplane spanned by S:

$$S = \{s = (s_1, \ldots, s_d) \in \mathbb{R}^d : s_i \geq 0, \sum_{i=1}^{d} s_i = 1\},$$
$$\langle S \rangle = \{s = (s_1, \ldots, s_d) \in \mathbb{R}^d : \sum_{i=1}^{d} s_i = 1\}.$$

Let $\mathrm{wt} : \mathbb{R}^d \to \mathbb{R}^k$ be the linear surjection that maps the i-th canonical basis vector of $\mathbb{R}^d$ to $a(i) \in \mathbb{R}^k$. For $w \in W(p)$, the inverse image $\mathrm{wt}^{-1}(w)$ is a translate of the kernel of wt; we will use the notations

$$\mathrm{wt}_S^{-1}(w) = \mathrm{wt}^{-1}(w) \cap S,$$
$$B_w(x, \delta) = B(x, \delta) \cap \mathrm{wt}^{-1}(w) \cap \langle S \rangle,$$

and denote by $\mathrm{int}_w(A)$ the interior of a set relative to $\mathrm{wt}^{-1}(w) \cap S$. The *entropy function* $h : S \to \mathbb{R}^+$ is defined by $h(s) = -\sum_{i=1}^{d} s_i \log(s_i)$. It is well-known (and easy to check) that h is strictly concave on S.

THEOREM 3.1. *(a) For each $w \in W(p)$, the restriction of h to $\mathrm{wt}_S^{-1}(w)$ achieves its maximum uniquely at a point $s(w) \in \mathrm{int}_w(\mathrm{wt}_S^{-1}(w))$.*

(b) The map $w \mapsto s(w)$ is C^1 on $W(p)$. More precisely, there exists a continuous map $w \mapsto L(w)$ of $W(p)$ into the space of linear endomorphisms of $\mathbb{R}^k$ such that, for all $\tilde{w} \in W(p)$ and nonzero $u \in \mathbb{R}^k$ with $\tilde{w} + \epsilon u \in W(p)$ for small $\epsilon > 0$, the corresponding directional derivative of $w \mapsto s(w)$ at $\tilde{w}$ is given by $L(\tilde{w})u/\|u\|$.

REMARK 3.2. The restriction of (3.1) to the relative interior of $W(p)$ is a special case of the results of [**MT2**]. To prove (3.1) on all of $W(p)$ we will import to ideas from section 2 of [**MT2**], and refine them in our setting. We remark that such an extension from the affine interior of the (weight-per-symbol) polytope to all of the polytope is not possible for the more general setting of [**MT2**]. (See [**CMT**].) We also remark that the map $w \mapsto s(w) : W(p) \to S$ of (3.1) is not in general much better than C^1; for instance, in the case of the polynomial $1+y^2+y^3$ in one variable $y = x_1$, it is not C^2.

We will denote the relative interior of $F \in \mathcal{F}(p)$ by $\text{int}(F)$. The fact that h achieves the maximum over $\text{wt}_S^{-1}(w)$ at a unique point $s(w) \in \text{wt}_S^{-1}(w)$ follows from the compactness of $\text{wt}_S^{-1}(w)$ and the strict concavity of h. If $w \in \text{int}(W(p))$ then we have $s(w) \in \text{int}_w(\text{wt}_S^{-1}(w))$ as a result of (8) of [**MT2**]. If $w \in \text{int}(F)$ for a proper face F of $W(p)$ then we deduce $s(w) \in \text{int}_w(\text{wt}_S^{-1}(w))$ by applying (8) of [**MT2**] to p_F. This establishes (3.1)(a). For the proof of (3.1)(b) we will adapt the description of $s(w)$ presented by [**MT2**].

For $j \in \{1, \ldots, k\}$, $x = (x_1, \ldots, x_k) \in \mathbb{R}^k$ and $w = (w_1, \ldots, w_k) \in \mathbb{R}^k$, let

$$f_j(x, w) = x_j \frac{\partial p}{\partial x_j} - w_j p = \sum_{a \in \text{Log}(p)} (a_j - w_j) p_a x^a,$$

and put $f = (f_1, \ldots, f_k)$ to obtain a function $f : \mathbb{R}^k \times \mathbb{R}^k \to \mathbb{R}^k$. It follows from (7) of [**MT2**] that for each $w \in \text{int}(W(p))$ there exists a unique $x = x(w) \in (\mathbb{R}^{++})^k$ such that

$$(3.1) \qquad f_j(x, w) = \sum_{a \in \text{Log}(p)} (a_j - w_j) p_a x^a = 0.$$

for all $j \in \{1, \ldots, k\}$. For $w \in \text{int}(W(p))$, (8) of [**MT2**] then shows that the i-th coordinate of $s(w) \in S$ is given by

$$(3.2) \qquad s(w)_i = \frac{x(w)^{a(i)}}{p(x(w))}.$$

In particular, if $w \in \text{int}(W(p))$ then $s(w) \in \text{int}(S)$, the relative interior of S. Let $D(x)$ denote the diagonal matrix with $x = (x_1, \ldots, x_k)$ as its diagonal and, for $w \in \text{int}(W(p))$, let $Q(w)$ be the $k \times k$ matrix with

$$Q(w)_{jj'} = \sum_{a \in \text{Log}(p)} (a_j - w_j)(a_{j'} - w_{j'}) p_a x(w)^a.$$

LEMMA 3.3. *For $w \in \text{int}(W(p))$ the matrix*

$$(Jf)(x(w), w) = \left(\frac{\partial f_j}{\partial x_{j'}} \right)_{1 \le j, j' \le k}$$

is invertible, and $(Jf)(x(w), w) = Q(w) D(x(w))^{-1}$.

PROOF. Let $w \in \text{int}(W(p))$ and $x = x(w)$. Multiplying the j'-th column of $(Jf)(x, w)$ by $x_{j'} > 0$,

$$(Jf)(x, w) D(x) = \left(\sum_{a \in \text{Log}(p)} (a_j - w_j) a_{j'} p_a x^a \right)_{1 \le j, j' \le k}.$$

Combining this with (3.1), we have $(Jf)(x, w) D(x) = Q(w)$. Define a probability vector π on $\text{Log}(p)$ by putting $\pi(a) = p_a x^a / p(x)$ for $a \in \text{Log}(p)$. For $j \in \{1, \ldots, k\}$, let g_j be the function defined on $\mathrm{Log}(p)$ by $g_j(a) = a_j - w_j$. Since $\mathrm{Log}(p) \subset \Delta(p) = \mathbb{Z}^k$, the column space of the $k \times |\mathrm{Log}(p)|$ matrix $(a_j - w_j)$ equals $\mathbb{R}^k$. Considering the row space, we see that the $|\mathrm{Log}(p)|$-vectors $g_1, \ldots, g_k$ are independent. Since $Q(w)$ is none other than the matrix of inner products

$$\langle g_j, g_{j'} \rangle = \int g_j\, g_{j'}\, d(p(x)\pi),$$

it is positive definite. So, $Q(w)$ and $(Jf)(x(w), w) = Q(w)D(x(w))^{-1}$ are non-singular. □

PROOF OF (3.1). We have already established (3.1)(a). Turning to (3.1)(b), first note that the equations (3.1) satisfied by $x(w)$, lemma (3.3), the implicit function theorem and (3.2) show that $w \mapsto x(w)$ and $w \mapsto s(w)$ are C^1 on $\mathrm{int}(W(p))$. We will handle $\partial W(p)$ by induction on the codimension of $F \in \mathcal{F}(p)$.

LEMMA 3.4. *If F is a facet of $W(p)$ and $\tilde{w} \in \mathrm{int}(F)$, then the map $w \mapsto s(w)$ is C^1 in a neighbourhood of $\tilde{w}$ in $W(p)$.*

PROOF. Using suitable changes of variables, it suffices to assume that

(i) $a(i)_k \geq 0$ for all $i \in \{1, \ldots, d\}$,
(ii) $a(i)_k = 0$ if and only if $a(i) \in F$,

and to verify that each coordinate $s(w)_i$ of $s(w)$ is C^1 in a neighbourhood of $\tilde{w}$ in $W(p)$. So, we assume (i) and (ii). Then the facet F is obtained from $W(p)$ by putting $w_k = 0$, and $\tilde{w}$ is of the form $\tilde{w} = (\tilde{w}_1, \ldots, \tilde{w}_{k-1}, 0)$. By (i), (ii) and applications of (7) and (8) of [**MT2**] to p_F it can be seen that, for $w \in \mathrm{int}(F)$ as well as for $w \in \mathrm{int}(W(p))$, the equations (3.1) have a unique solution $x(w) \in (\mathbb{R}^+)^k$ and $s(w)$ is given by (3.2). Note also that for $w \in \mathrm{int}(F)$ we have $x(w)_k = 0$ and $x(w)_1, \ldots, x(w)_{k-1} > 0$. (We remark that $x(w)$ need not be differentiable at $\tilde{w}$.) Let

$$m = \min\{a(i)_k : 1 \leq i \leq d,\, a(i)_k > 0\}.$$

Picking suitable $d_0, d_1 \in \{1, \ldots, d\}$, order the monomials $a(1), \ldots, a(d)$ so that $a(i)_k = 0$ for $i = 1, \ldots, d_0$; $a(i)_k = m$ for $i = d_0 + 1, \ldots, d_1$; and $a(i)_k > m$ for $i = d_1 + 1, \ldots, d$. Let

$$p_{F,m}(x_1, \ldots, x_{k-1}) = m \sum_{i=d_0+1}^{d_1} x_1^{a(i)_1} \cdots x_{k-1}^{a(i)_{k-1}}.$$

For $j = k$ and small $w_k > 0$, evaluate the equation (3.1) at $w = (\tilde{w}_1, \ldots, \tilde{w}_{k-1}, w_k)$ to obtain

$$\sum_{i=d_0+1}^{d_1} (m - w_k) x(w)^{a(i)} + \sum_{i=d_1+1}^{d} (a(i)_k - w_k) x(w)^{a(i)} = \sum_{i=1}^{d_0} w_k\, x(w)^{a(i)}.$$

Factor $(x(w)_k)^m$ out of the left-hand side, divide the equation by w_k and let $w_k \to 0$ to see that

$$\lim_{w_k \to 0^+} \frac{(x(w)_k)^m}{w_k} = p_F(x(\tilde{w}))/p_{F,m}(x(\tilde{w})).$$

It follows that, for $a_k \geq m$, the directional derivative of $x(w)^a$ in the direction of w_k exists at $\widetilde{w}$ and equals

(3.3) $$x(\widetilde{w})_1^{a_1} \cdots x(\widetilde{w})_{k-1}^{a_{k-1}} 0^{a_k - m} p_F(x(\widetilde{w})) / p_{F,m}(x(\widetilde{w})).$$

Since $x(w)_k = 0$ on $\operatorname{int}(F)$, for $a_k \geq 1$ the partial derivatives of $x(w)^a$ with respect to $w_1, \ldots, w_{k-1}$ equal 0. These derivatives are clearly continuous on $\operatorname{int}(F)$. To check continuity in a neighbourhood of $\widetilde{w}$ in $W(p)$, consider $w = (w_1, \ldots, w_k) \in \operatorname{int}(W(p))$. Let $\left(\frac{\partial x}{\partial w}\right)$ denote the $k \times k$ matrix with

$$\left(\frac{\partial x}{\partial w}\right)_{ij} = \frac{\partial (x(w)_i)}{\partial w_j}.$$

As a result of (3.1), the matrix product

$$(Jf)(x(w), w) \left(\frac{\partial x}{\partial w}\right) = -pI.$$

Using (3.3) to solve for $\left(\frac{\partial x}{\partial w}\right)$,

$$\frac{\partial (x(w)_i)}{\partial w_j} = -p\, x(w)_i\, (Q(w)^{-1})_{ij} = -p\, x(w)_i\, \operatorname{Adj}(Q(w))_{ij} / \det(Q(w)).$$

from this one explicitly computes $\frac{\partial (x(w)^a)}{\partial w_j}$ for each $a \in \operatorname{Log}(p)$ with $a_k \geq m$ and takes the limit as $w \to \widetilde{w}$ to verify that the limit equals 0 for $j = 1, \ldots, k-1$ and (3.3) for $j = k$. This proves that for $i = d_0 + 1, \ldots, d$ the function $w \mapsto x(w)^{a(i)}$ is C^1 in a neighbourhood of $\widetilde{w}$ in $W(p)$.

Now consider, for $j = 1, \ldots, k-1$, the C^1 function

$$g_j(w) = \sum_{i=d_0+1}^{d} (a(i)_j - w_j)\, x(w)^{a(i)}$$

and note that the equations (3.1) may be expressed as

$$\sum_{i=1}^{d_0} (a(i)_j - w_j)\, x^{a(i)} + g_j(w) = 0.$$

The $(k-1) \times (k-1)$ matrix of the partial derivatives of the left-hand sides with respect to $x_1, \ldots, x_{k-1}$ has

$$\sum_{a \in \operatorname{Log}(p_F)} (a_j - w_j)\, a_{j'}\, p_a\, x^a / x_{j'}$$

for its (j, j') entry, and an application of (3.3) to p_F reveals that this matrix is invertible. So, we can apply the implicit function theorem to deduce that $x(w)_1, \ldots, x(w)_{k-1}$ are C^1 in a neighbourhood of $\widetilde{w}$ in $W(p)$. Hence $x^{a(i)}$ is C^1 for $i = 1, \ldots, d$ and, by (3.2), so is $s(w)$. □

To complete the proof of (3.1) by induction, consider $F \in \mathcal{F}(p)$ with $\dim(F) < k - 1$, assuming that (3.1)(b) has been verified when $\widetilde{w} \in \operatorname{int}(G)$ for $G \in \mathcal{F}(p)$ with

$\dim(G) > \dim(F)$. Suppose $\tilde{w} \in \operatorname{int}(F)$. Let V denote the nonzero elements of

$$(F - F) \cup \bigcup_{\substack{G \in \mathcal{F}(p),\\ \dim(G)=\dim(F)+1}} (G - F).$$

Apply (3.4) to p_G for each $G \in \mathcal{F}(p)$ with $\dim(G) = \dim(F) + 1$ to see that the directional derivative of $s(w)$ in the direction of v exists for every $v \in V$. Then, as in the proof of (3.4), use (3.3) to verify the continuity of these derivatives on $B(\tilde{w}, \delta) \cap \operatorname{int}(W(p))$ for small $\delta > 0$. Taken with our inductive assumption, this shows the continuity of these directional derivatives on a neighbourhood of $\tilde{w}$ in $W(p)$. As nonnegative combinations of vectors in V yield all u such that $\tilde{w} + \epsilon u \in W(p)$ for small $\epsilon > 0$, it follows that $s(w)$ is C^1 on a neighbourhood of $\tilde{w}$ in $W(p)$. Since $W(p) = \bigcup_{F \in \mathcal{F}(p)} \operatorname{int}(F)$, this proves (3.1). $\square$

4

Equilibrium distributions and coefficients of p^n

We now fix $F \in \mathcal{F}(p)$, and let $\bar{\epsilon}_F > 0$. In addition to the notation and assumptions set up at the beginning of section 3, we assume that $a(1), \ldots, a(d)$ have been ordered so that $a(1), \ldots, a(d_0) \in F$ and $a(d_0 + 1), \ldots, a(d) \in \text{Log}(p) \backslash F$.

In the sequel G will denote a set of the form

$$G = \overline{B(F, \epsilon_F) \backslash B(\partial F, \bar{\epsilon}_F)},$$

where $\epsilon_F > 0$ is small enough for G to be bounded away from $\partial W(p) \backslash \text{int}(F)$. (The proofs of (1.2)–(1.4) will place further requirements on ϵ_F.) The first four lemmas below follow from the compactness of G, the fact that G is bounded away from $\partial W(p) \backslash \text{int}(F)$, and the continuity of $s(w)$ on $W(p)$.

LEMMA 4.1. *There exists $\delta_1 > 0$ such that*
  (i) $B_w(s(w), \delta_1) \subset \text{int}_w(\text{wt}_S^{-1}(w))$ *for all $w \in G$;*
  (ii) *s_i is bounded away from 0 for $s \in \bigcup_{w \in G} B_w(s(w), \delta_1)$ and $i = 1, \ldots, d_0$.*

LEMMA 4.2. *Let $\delta_1 > 0$. For any $\delta_2 \in (0, \delta_1)$ there exists $\xi > 0$ such that*

$$B_{w+\text{wt}(t)}(s(w + \text{wt}(t)), \delta_2) \subset B_w(s(w), \delta_1) + t$$

whenever $w \in G$, $t \in \mathbb{R}^d$ and $\|t\| < \xi$.

LEMMA 4.3. *For $\delta_2 > 0$ there exists $\eta > 0$ such that*

$$h(s(w)) - \eta > h(s)$$

whenever $w \in G$ and $s \in \text{wt}_S^{-1}(w) \backslash B_w(s(w), \delta_2)$.

LEMMA 4.4. *For $\delta_2, \eta > 0$ there exists $\delta_3 \in (0, \delta_2)$ such that $h(s) > h(s(w)) - \eta/2$ whenever $w \in G$ and $s \in B_w(s(w), \delta_3)$.*

For the sequel, let $\delta_1 > 0$ satisfy (4.1).

LEMMA 4.5. *There exists $K_1 > 0$ such that for any $b \in \Delta(p)$ we have $\bar{b} = (\bar{b}_1, \ldots, \bar{b}_d) \in \mathbb{Z}^d$ such that*
  (i) $\text{wt}(\bar{b}) = b$,
  (ii) $\sum_{i=1}^{d} \bar{b}_i = 0$,
  (iii) $\|\bar{b}\| \leq K_1 \|b\|$.

If $b \in \Delta(p_F)$ then $\bar{b}$ can be chosen so that $\bar{b}_i = 0$ for $i = d_0 + 1, \ldots, d$.

PROOF. For $j \in \{1, \ldots, k\}$, the j-th canonical basis vector $e(j)$ of $\mathbb{R}^k$ belongs to $\mathbb{Z}^k = \Delta(p)$. By the definition of $\Delta(p)$, we can find $v(j) \in \mathbb{Z}^d$ such that $\operatorname{wt}(v(j)) = e(j)$ and $\sum_{i=1}^d v(j)_i = 0$. Then $K_1 = \sum_{j=1}^k \|v(j)\|$ will do the trick. The final statement of the lemma is a consequence of the definition of $\Delta(p_F)$. □

LEMMA 4.6. *Let $\delta_3 \in (0, \delta_1)$. There exists $N_1 \in \mathbb{N}$ such that $nB_w(s(w), \delta_3) \cap (\mathbb{Z}^+)^d \neq \emptyset$ whenever $n \geq N_1$, $w \in G$ and $(p^n)_{nw} \neq 0$. Moreover, there is a constant τ and a function $(w, n) \mapsto s(w, n)$ defined on $\mathcal{E} \equiv \{(w, n) : w \in G, n \geq N_1, (p^n)_{nw} \neq 0\}$ such that $s(w, n) \in nB_w(s(w), \delta_3) \cap (\mathbb{Z}^+)^d$ and $\|s(w, n) - s(w', n)\| \leq n\tau \|w - w'\| + \tau$ whenever $(w, n), (w', n) \in \mathcal{E}$.*

PROOF. The proof is a modification of lemma 15 of [**MT2**] as follows. Let $w \in G$ and $n \in \mathbb{N}$. Start by approximating $ns(w)$ by an integral vector t: Writing $s = s(w)$, define t by putting $t_i = \lfloor ns_i \rfloor$ for $i = 1, \ldots, d-1$ and

$$t_d = n - \sum_{i=1}^{d-1} t_i.$$

Note that t_i differs from ns_i by at most 1 for $i = 1, \ldots, d-1$. It follows from this and the fact that $ns_d = n - \sum_{i=1}^{d-1} ns_i$ that t_d differs from ns_d by at most $d-1$. So, $\|ns - t\|$ is bounded above by a constant K_2 that depends only on d. Now suppose $(p^n)_{nw} \neq 0$. Then $nw \in \mathbb{Z}^k$, so that $nw - \operatorname{wt}(t) \in \Delta(p) = \mathbb{Z}^k$. Applying (4.5), we have $\beta \in \mathbb{Z}^d$ such that $\operatorname{wt}(\beta) = nw - \operatorname{wt}(t)$, $\sum_{i=1}^d \beta_i = 0$ and

$$\|\beta\| \leq K_1 \|nw - \operatorname{wt}(t)\| = K_1 \|\operatorname{wt}(ns(w) - t)\| \leq K_1 \|\operatorname{wt}\| K_2,$$

where K_1 is as in (4.5) and $\|\operatorname{wt}\|$ is the norm of the linear map $s \mapsto \operatorname{wt}(s)$. Let $r(w) = t + \beta$. Then

$$\|r(w) - ns(w)\| \leq \|t - ns(w)\| + \|\beta\| \leq K_2 + K_1 \|\operatorname{wt}\| K_2.$$

Since the upper bound is independent of $w \in G$, there is $N_1 \in \mathbb{N}$ such that $r(w) \in nB(s(w), \delta_3) \cap \mathbb{Z}^d$ for all $n \geq N_1$ and $w \in G$. It remains to check that $r = r(w)$ is such that $\frac{r}{n} \in \operatorname{wt}_S^{-1}(w) = \operatorname{wt}^{-1}(w) \cap S$. Since $\operatorname{wt}(r) = \operatorname{wt}(t) + \operatorname{wt}(\beta) = nw$, we have $\frac{r}{n} \in \operatorname{wt}^{-1}(w)$. Since the vector t sums to n and β sums to 0, the vector $\frac{r}{n}$ sums to 1. Combining these observations with (4.1) and the fact that $\delta_3 < \delta_1$,

$$\frac{r}{n} \in B(s(w), \delta_1) \cap \operatorname{wt}^{-1}(w) \cap \langle S \rangle = B_w(s(w), \delta_1) \subset \operatorname{int}_w(\operatorname{wt}_S^{-1}(w)).$$

In particular, $\frac{r}{n} \in S$. Take $s(w, n) = r(w)$.

For the second claim of the lemma, consider $s(w, n) = t + \beta$ and, for $(w', n) \in \mathcal{E}$, denote the corresponding objects by t', β' and $s(w', n) = t' + \beta'$. We have

$$\|s(w, n) - s(w', n)\|$$
$$\leq \|t - t'\| + \|\beta - \beta'\|$$
$$\leq \|t - ns(w)\| + \|t' - ns(w')\| + n\|s(w) - s(w')\| + \|\beta\| + \|\beta'\|.$$

In the latter sum, the first two terms are bounded above by K_2; the third term is bounded above by $nM\|w - w'\|$ where M is a constant resulting from (3.1)(b); and each of the last two terms is bounded above by $K_1 \|\operatorname{wt}\| K_2$. Hence,

$$\|s(n, w) - s(n, w')\| \leq nM\|w - w'\| + 2K_2 + 2K_1 K_2 \|\operatorname{wt}\|. \quad \square$$

Let
$$T(w,n) = (n\,\mathrm{wt}_S^{-1}(w)) \cap \mathbb{Z}^d.$$
Note that each $t = (t_1,\ldots,t_d) \in T(w,n)$ has $t_i \in \mathbb{Z}^+$ and $\sum_{i=1}^d t_i = n$. Hence
$$|T(w,n)| \le |[0,n] \cap \mathbb{Z}|^d = (n+1)^d.$$
Each element of $T(w,n)$ represents a selection of n elements of $\mathrm{Log}(p)$ which sum to $nw \in \mathrm{Log}(p^n)$: Putting
$$\alpha(t) = n!/(t_1!\cdots t_d!),$$
we have
$$(p^n)_{nw} = \sum_{t \in T(w,n)} \alpha(t).$$
It is easily deduced from Stirling's formula (as on p. 250 of [**MT2**]) that there is a constant K_3 so that, the entropy $h = h\left(\frac{1}{n}t\right)$ and $\alpha(t)$ satisfy
$$(4.1) \qquad K_3^{-1} n^{-d} e^{nh} \le \alpha(t) \le K_3 n e^{nh}.$$
Let $\delta_2 > 0$. Choose δ_3, η to satisfy (4.3) and (4.4). Decompose $(p^n)_{nw}$ as
$$(p^n)_{nw} = \sum_{t \in T(w,n) \cap nB_w(s(w),\delta_2)} \alpha(t) + \sum_{t \in T(w,n) \setminus nB_w(s(w),\delta_2)} \alpha(t).$$
If $nw \in \mathrm{Log}(p^n)$ then, by (4.6), the first sum is non-vacuous since it contains contributions from $s(w,n) \in T(w,n) \cap nB_w(s(w),\delta_3)$. By lemma (4.4) and the lower bound in (4.1) we see that the first sum is at least
$$K_3^{-1} n^{-d} e^{n(h(s(w))-\eta/2)}.$$
The second sum contains at most $|T(w,n)| \le (n+1)^d$ terms and, by (4.3) and the upper bound in (4.1), each of these terms is less than $K_3\, n\, e^{n(h(s(w))-\eta)}$. This proves the following.

THEOREM 4.7. *For $\delta_2 > 0$,*
$$\lim_{\substack{n \to \infty, \\ nw \in \mathrm{Log}(p^n)}} \sum_{t \in T(w,n) \cap nB_w(s(w),\delta_2)} \alpha(t)/(p^n)_{nw} = 1$$
uniformly over $w \in G$.

5

Proofs of the estimates

We retain the notation and assumptions set up at the beginnings of sections 3 and 4.

PROOF OF (1.2). Put $M = 2L\max\{\|u\| : u \in \text{Log}(p)\}$. Let $\delta_0 > 0$ be such that for all $w \in F\backslash B(\partial F, \bar{\epsilon}_F)$ and $i, j \in \{1, \ldots, d_0\}$ we have

$$s(w)_i/s(w)_j > \delta_0.$$

Then for $\epsilon_F > 0$ sufficiently small, $w \in B(F, \epsilon_F)\backslash B(\partial F, \bar{\epsilon}_F)$ and $i, j \in \{1, \ldots, d_0\}$ we have

(5.1) $$s(w)_i/s(w)_j > \delta_0/2.$$

Thus, for any constant $K_4 > 0$, we can choose $\epsilon_F > 0$ so small that for $w \in B(F, \epsilon_F)\backslash B(\partial F, \bar{\epsilon}_F)$, positive δ_1 satisfying (4.1), all large n and $t \in nB_w(s(w), \delta_1)$ we have

(5.2) $$(t_i - M)/(t_j + M) > \delta_0/3 \text{ whenever } i, j \in \{1, \ldots, d_0\}$$

and

(5.3) $$(t_i - M)/(t_j + M) > K_4 \text{ whenever } i \in \{1, \ldots, d_0\}, j \in \{d_0 + 1, \ldots d\}.$$

This is possible by (5.1) and the fact that, as ϵ_F decreases to 0, the sum $\sum_{i=1}^{d_0} s(w)_i$ tends to 1 uniformly for $w \in B(F, \epsilon_F)\backslash B(\partial F, \bar{\epsilon}_F)$.

For small $\epsilon_F > 0$, we put $G = B(F, 2\epsilon_F)\backslash B(\partial F, \bar{\epsilon}_F/2)$ and

$$G_0 = B(F, \epsilon_F)\backslash B(\partial F, \bar{\epsilon}_F).$$

Note that $G_0 \subset G$. Let $\delta_1 > 0$ satisfy (4.1) for our choice of G. Fixing $\delta_2 \in (0, \delta_1)$, let $\xi > 0$ be as given by (4.2) for our G. For $b \in (\text{Log}(p^L)\backslash \text{Log}(p_F^L)) - \text{Log}(p_F^L)$, find $\bar{b} \in (\mathbb{Z}^+)^d$ such that

(i) $\text{wt}(\bar{b}) = b$,
(ii) $\sum_{i=1}^{d} \bar{b}_i = 0$,
(iii) $\|\bar{b}\| \leq M$,
(iv) $\bar{b}_i \geq 0$ for $i = d_0 + 1, \ldots, d$,
(v) $\sum_{i=d_0+1}^{d} \bar{b}_i > 0$.

Considering $a \in \mathbb{Z}^k$ and $n \in \mathbb{N}$ such that $w = \frac{a}{n} \in G_0$, we use two requirements to specify N so that $(p^n)_{a+b} \geq K(p^n)_a$ if $n \geq N$. The first requirement is that $M/N < \xi$. Then, for $n > N$, lemma (4.2) applies to $w = a/n$ and $t = \bar{b}/n$ and ensures that $(T(w, n) \cap nB_w(s(w), \delta_1)) + \bar{b}$ contains $T(w + b/n, n) \cap nB_{w+b/n}(s(w + b/n), \delta_2)$.

Put $U = T(w,n) \cap nB_w(s(w), \delta_1)$. We also require N to be so that for $n \geq N$ we have $w + b/n \in G$ and, applying (4.7) on G,
$$(p^n)_{a+b}/(p^n)_a$$
is within $K/2$ of
$$\frac{\sum_{t \in U} \alpha(t + \bar{b})}{\sum_{t \in U} \alpha(t)}.$$
Consequently, to prove the proposition it suffices to show that
$$\alpha(t + \bar{b})/\alpha(t)$$
is at least $3K/2$ for each $t \in U$. We write $\{1, \ldots, d_0\} = V_+ \cup V_-$ where V_+ is the subset of all $i \in \{1, \ldots, d_0\}$ with $\bar{b}_i > 0$ and V_- is the subset of all $i \in \{1, \ldots, d_0\}$ with $\bar{b}_i \leq 0$. Then
$$\alpha(t + \bar{b})/\alpha(t) = d_1/(d_2 d_3),$$
where
$$d_1 = \prod_{i \in V_-} (t_i + \bar{b}_i + 1) \cdots t_i,$$
$$d_2 = \prod_{i \in V_+} (t_i + 1) \cdots (t_i + \bar{b}_i),$$
$$d_3 = \prod_{i = d_0 + 1}^{d} (t_i + 1) \cdots (t_i + \bar{b}_i).$$
Put $\beta_1 = -\sum_{i \in V_-} \bar{b}_i$, $\beta_2 = \sum_{i \in V_+} \bar{b}_i$, $\beta_3 = \sum_{i = d_0 + 1}^{d} \bar{b}_i$ and
$m_1 = \min\{t_i : i \in V_-\}$, $m_2 = \max\{t_i : i \in V_+\}$, $m_3 = \max\{t_i : d_0 + 1 \leq i \leq d\}$.
Then
$$d_1 \geq (m_1 - M)^{\beta_1},$$
$$d_2 \leq (m_2 + M)^{\beta_2},$$
$$d_3 \leq (m_3 + M)^{\beta_3}.$$
Using (5.2), (5.3) and the fact that $\beta_1 = \beta_2 + \beta_3$, we find that
$$\frac{d_1}{d_2 d_3} \geq \left(\frac{m_1 - M}{m_2 + M}\right)^{\beta_2} \left(\frac{m_1 - M}{m_3 + M}\right)^{\beta_3} \geq \left(\frac{\delta_0}{3}\right)^{\beta_2} (K_4)^{\beta_3}.$$
Since $\beta_3 > 0$ by (v) of our choice of $\bar{b}$, the last quantity will be greater than $3K/2$ for sufficiently large K_4 (which entails choosing $\epsilon_F > 0$ sufficiently small). □

PROOF OF (1.3). Let $\bar{\epsilon}_F, D$ be as in (1.3). Pick $\epsilon_F > 0$ such that
$$G = \overline{B(F, 2\epsilon_F) \backslash B(\partial F, \bar{\epsilon}_F/2)}$$
is bounded away from $\partial W(p) \backslash \text{int}(F)$. We will find N such that (1.3) holds for $n \geq N$ and $a, a + u \in \text{Log}(p^n)$, $b \in \Delta(p_F)$ with $w = \frac{a}{n} \in G_0 = \overline{B(F, \epsilon_F) \backslash B(\partial F, \bar{\epsilon}_F)}$ and $\|b\|, \|u\| \leq D$. Let $\delta_1 > 0$ satisfy (4.1) for G, and find $\delta_0 > 0$ such that

$s_i > \delta_0$ for $s \in \bigcup_{w \in G} B_w(sw), \delta_1)$ and $i \in \{1, \ldots, d_0\}$. Since $G_0 \subset G$, then δ_1 or any smaller positive number will satisfy (4.1) for both G_0 and G (and δ_0 will continue to serve as the lower bound). Let K_1 be as in (4.5) and decrease δ_1 so that $0 < \delta_1 < \delta_0$ and

(5.4)
$$\left(\frac{\delta_0 + \delta_1}{\delta_0 - \delta_1}\right)^{(K_1 D)^{d_0}} \leq \min\left\{\frac{4}{4-\theta}, \frac{4+\theta}{4}\right\}.$$

Picking $0 < \delta_3 < \delta_2 < \delta_1$, let ξ and τ be as in (4.2) and (4.6) for our choice of G. Put
$$M = \max\{K_1 D, \tau D + \tau\}.$$

Apply (4.5) to $b \in \Delta(p_F)$ to pick $\bar{b} \in \mathbb{Z}^d$ supported on coordinates $1, \ldots, d_0$ such that $\mathrm{wt}(\bar{b}) = b$, $\sum_{i=1}^d \bar{b}_i = 0$ and
$$\|\bar{b}\| \leq K_1 \|b\| \leq K_1 D \leq M.$$

Make sure N is large enough to have $2M/N < \xi$ and for (4.6) to apply to $n \geq N$ to provide $s(w, n) \in T(w, n) \cap nB_w(s(w), \delta_3)$ and $s(w + \frac{u}{n}) \in T(w + u/n, n) \cap nB_{w+u/n}(s(w+u/n), \delta_3)$ with
$$\|s(w + \frac{u}{n}) - s(w, n)\| \leq \tau\|u\| + \tau \leq \tau D + \tau \leq M.$$

Put $\bar{u} = s(w + \frac{u}{n}) - s(w, n)$. Since $\|\bar{u}/n\|, \|\bar{b}/n\|, \|(\bar{u}+\bar{b})/n\| < \xi$ and $\bar{u}, \bar{b}$ both sum to 0, we obtain from (4.2) the inclusions

$$T(w + b/n, n) \cap nB_{w+\frac{b}{n}}(s(w + b/n), \delta_2) \subset (T(w, n) \cap nB_w(s(w), \delta_1)) + \bar{b},$$

$$T(w + u/n, n) \cap nB_{w+\frac{u}{n}}(s(w + u/n), \delta_2) \subset (T(w, n) \cap nB_w(s(w), \delta_1)) + \bar{u},$$

as well as the fact that $T(w + u/n + b/n, n) \cap nB_{w + \frac{u+b}{n}}(s(w + u/n + b/n), \delta_2)$ is contained in $(T(w, n) \cap nB_w(s(w), \delta_1)) + \bar{u} + \bar{b}$. Let $U = T(w, n) \cap nB_w(s(w), \delta_1)$. Applying (4.7) to our choice of G, we see that for sufficiently large N and $n \geq N$

$$\frac{(p^n)_a / (p^n)_{a+b}}{(p^n)_{a+u} / (p^n)_{a+u+b}}$$

is within $\theta/2$ of

$$\frac{\left(\sum_{t \in U} \alpha(t)\right)\left(\sum_{t \in U} \alpha(t + \bar{u} + \bar{b})\right)}{\left(\sum_{t \in U} \alpha(t + \bar{b})\right)\left(\sum_{t \in U} \alpha(t + \bar{u})\right)} = \frac{\sum_{t, t' \in U} \alpha(t)\, \alpha(t' + \bar{u} + \bar{b})}{\sum_{t, t' \in U} \alpha(t + \bar{b})\, \alpha(t' + \bar{u})}.$$

So, to complete the proof it suffices to ensure that

(5.5)
$$\frac{\alpha(t)\, \alpha(t' + \bar{u} + \bar{b})}{\alpha(t + \bar{b})\, \alpha(t' + \bar{u})}$$

is within $\theta/2$ of 1 for all $t, t' \in U$. Let V_+ be the set of $i \in \{1, \ldots, d_0\}$ with $\bar{b}_i > 0$ and V_- the set of $i \in \{1, \ldots, d_0\}$ with $\bar{b}_i < 0$. Since $\bar{b}_i = 0$ for $i = d_0 + 1, \ldots, d$, the ratio (5.5) equals

$$\prod_{i\in V_+} \frac{(t_i+1)\cdots(t_i+\bar{b}_i)}{(t'_i+\bar{u}_i+1)\cdots(t'_i+\bar{u}_i+\bar{b}_i)} \cdot \prod_{i\in V_-} \frac{(t'_i+\bar{u}_i+\bar{b}_i+1)\cdots(t'_i+\bar{u}_i)}{(t_i+\bar{b}_i+1)\cdots t_i}.$$

This can be viewed as the product of at most $(K_1 D)^{d_0}$ terms, each of which equals, or is the reciprocal of, a ratio

$$\frac{t_i+\mu+\beta}{t'_i+\mu} = \frac{1+(\mu+\beta)/t_i}{(t'_i/t_i)+(\mu/t_i)}$$

in which $i \in \{1,\ldots,d_0\}$ and $|\mu|, |\beta| \le M$. Since $t/n \in B_w(s(w), \delta_1)$, we have $t_i \ge n\delta_0$. So, the ratios $(\mu+\beta)/t_i$ and μ/t_i can be made arbitrarily small by picking N sufficiently large. The proof is completed by observing

$$\frac{\delta_0-\delta_1}{\delta_0+\delta_1} \le \frac{s(w)_i-\delta_1}{s(w)_i+\delta_1} \le \frac{t'_i}{t_i} = \frac{t'_i/n}{t_i/n} \le \frac{s(w)_i+\delta_1}{s(w)_i-\delta_1} \le \frac{\delta_0+\delta_1}{\delta_0-\delta_1}$$

and recalling the requirement (5.4) we made of δ_1. □

PROOF OF (1.4). It is enough to prove the proposition for $l = 1$. By the definition of c_F, there exists $\bar{c} \in \mathbb{Z}^d$ such that $\text{wt}(\bar{c}) = c_F$, $\sum_{i=1}^d \bar{c}_i = 1$ and $\bar{c}$ is supported only on coordinates $1, \ldots, d_0$. For $b \in \Delta(p_F)$ with $\|b\| \le D$, apply (4.5) to find $\bar{b} \in \mathbb{Z}^d$ such that $\text{wt}(\bar{b}) = b$, $\|\bar{b}\| \le K_1 D$, $\sum_{i=1}^d \bar{b}_i = 0$ and $\bar{b}$ is supported on coordinates $1, \ldots, d_0$. Let

$$M = K_1 D + \|\bar{c}\|.$$

Choose $K_5 > 0$ such that for all $w \in F\setminus B(\partial F, \bar{\epsilon}_F)$ and $i, j \in \{1, \ldots, d_0\}$

$$\frac{1}{K_5} < s(w)_i/s(w)_j < K_5.$$

Let $\epsilon_F > 0$ be such that $G \equiv \overline{B(F, 2\epsilon_F)\setminus B(\partial F, \bar{\epsilon}_F/2)}$ is bounded away from $\partial W(p)\setminus \text{int}(F)$, and let $\delta_1 > 0$ be as in (4.1) for this choice of G. Put

$$G_0 = \overline{B(F, \epsilon_F)\setminus B(\partial F, \bar{\epsilon}_F)}.$$

By decreasing the positive numbers ϵ_F, δ_1 and taking N to be sufficiently large, we make sure that for $w \in G_0$, $n \ge N$, $t \in T(w,n) \cap nB_w(s(w), \delta_1)$ and $i, j \in \{1, \ldots, d_0\}$ we have

(5.6) $$\frac{1}{2K_5} < \frac{t_i}{t_j} < \frac{t_i+M}{t_j-M} < 2K_5$$

and

(5.7) $$\frac{1}{2d_0 K_5} < \frac{n+1}{t_i+M} < \frac{n+1}{t_i-M} < 2d_0 K_5.$$

Let $0 < \delta_2 < \delta_1$ and choose K such that

$$K - \frac{1}{2K} > (2d_0 K_5)(2K_5)^M$$

and

$$\frac{3}{2K} < \frac{1}{2d_0 K_5 (2K_5)^M}.$$

Suppose $a \in \mathbb{Z}^k$ is such that $w = \frac{a}{n} \in G_0$. Provided $M/N < \xi$ and $n \geq N$, we can apply (4.2) on G to have
$$T(w + c_F/n + b/n, n) \cap nB_{w + \frac{c_F}{n} + \frac{b}{n}}(s(w + c_F/n + b/n), \delta_2)$$
contained in $(T(w, n) \cap nB(s(w), \delta_1)) + \bar{c} + \bar{b}$. Let $U = T(w, n) \cap nB(s(w), \delta_1)$. Applying (4.7) to G, we see that for sufficiently large N and $n \geq N$
$$(p^{n+1})_{a+c_F+b}/(p^n)_a$$
is within $\frac{1}{2K}$ of
$$\frac{\sum_{t \in U} \alpha(t + \bar{c} + \bar{b})}{\sum_{t \in U} \alpha(t)}.$$
So, it will suffice to show that
$$\alpha(t + \bar{c} + \bar{b})/\alpha(t)$$
is between $\frac{3}{2K}$ and $K - \frac{1}{2K}$ for each $t \in U$. Let V_+ be the set of $i \in \{1, \ldots, d_0\}$ such that $\bar{c}_i + \bar{b}_i > 0$, and let V_- be the set of $i \in \{1, \ldots, d_0\}$ with $\bar{c}_i + \bar{b}_i < 0$. Then
$$\alpha(t + \bar{c} + \bar{b})/\alpha(t) = (n+1)d_1/d_2,$$
where
$$d_1 = \prod_{i \in V_-} (t_i + \bar{c}_i + \bar{b}_i + 1) \cdots t_i,$$
$$d_2 = \prod_{i \in V_+} (t_i + 1) \cdots (t_i + \bar{c}_i + \bar{b}_i).$$
Put $\beta_1 = -\sum_{i \in V_-} \bar{c}_i + \bar{b}_i$, $\beta_2 = \sum_{i \in V_+} \bar{c}_i + \bar{b}_i$ and $m_1 = \min\{t_i : i \in V_-\}$, $M_1 = \max\{t_i : i \in V_-\}$, $m_2 = \min\{t_i : i \in V_+\}$, $M_2 = \max\{t_i : i \in V_+\}$. Then
$$(m_1 - \|\bar{c}\| - K_1 D)^{\beta_1} \leq d_1 \leq M_1^{\beta_1}$$
and
$$m_2^{\beta_2} \leq d_2 \leq (M_2 + \|\bar{c}\| + K_1 D)^{\beta_2}.$$
Observing that $\beta_2 - \beta_1 = \sum_{i=1}^{d} \bar{c}_i + \sum_{i=1}^{d} \bar{b}_i = 1$,
$$\frac{n+1}{M_2 + M} \left(\frac{m_1 - M}{M_2 + M}\right)^{\beta_1} \leq \frac{(n+1)d_1}{d_2} \leq \frac{n+1}{m_2} \left(\frac{M_1}{m_2}\right)^{\beta_1}.$$
By (5.6) and (5.7), the upper bound is at most $(2d_0 K_5)(2K_5)^{\beta_1}$ and the lower bound at least $(2d_0 K_5(2K_5)^{\beta_1})^{-1}$. By our choice of K, it follows that for each $t \in U$
$$\alpha(t + \bar{c} + \bar{b})/\alpha(t) = (n+1)d_1/d_2$$
is indeed between $\frac{3}{2K}$ and $K - \frac{1}{2K}$. □

Bibliography

[CMT] E. Cawley, B. Marcus and S. Tuncel, Boundary measures of Markov chains, *Israel J. Math.* **94** (1996), 111–123.

[H1] D. Handelman, Positive polynomials and product type actions of compact groups, *Mem. Amer. Soc.* **320** (1985).

[H2] D. Handelman, Deciding eventual positivity of polynomials, *Ergod. Th. and Dynam. Sys.* **6** (1986), 57–79.

[MT1] B. Marcus and S. Tuncel, The weight-per-symbol polytope and scaffolds of invariants associated with Markov chains, *Ergod. Th. and Dynam. Sys.* **11** (1991), 129–180.

[MT2] B. Marcus and S. Tuncel, Entropy at a weight-per-symbol and embeddings of Markov chains, *Invent. Math.* **102** (1990), 235–266.

[MT3] B. Marcus and S. Tuncel, Matrices of polynomials, positivity, and finite equivalence of Markov chains, *J. Amer. Math. Soc.* **6** (1993), 131–147.

[MT4] B. Marcus and S. Tuncel, Resolving Markov chains onto Bernoulli shifts, in this issue of *Mem. Amer. Math. Soc.*

[PS] W. Parry and K. Schmidt, Natural coefficients and invariants for Markov shifts, *Invent. Math.* **76** (1984), 15–32.

[T] S. Tuncel, Faces of Markov chains and matrices of polynomials, *Contemp. Math.*, Vol. 135, Amer. Math. Soc., Providence, 1992, pp. 391–422.

Editorial Information

To be published in the *Memoirs*, a paper must be correct, new, nontrivial, and significant. Further, it must be well written and of interest to a substantial number of mathematicians. Piecemeal results, such as an inconclusive step toward an unproved major theorem or a minor variation on a known result, are in general not acceptable for publication. Papers appearing in *Memoirs* are generally longer than those appearing in *Transactions*, which shares the same editorial committee.

As of November 30, 2000, the backlog for this journal was approximately 10 volumes. This estimate is the result of dividing the number of manuscripts for this journal in the Providence office that have not yet gone to the printer on the above date by the average number of monographs per volume over the previous twelve months, reduced by the number of volumes published in four months (the time necessary for preparing a volume for the printer). (There are 6 volumes per year, each containing at least 4 numbers.)

A Consent to Publish and Copyright Agreement is required before a paper will be published in the *Memoirs*. After a paper is accepted for publication, the Providence office will send a Consent to Publish and Copyright Agreement to all authors of the paper. By submitting a paper to the *Memoirs*, authors certify that the results have not been submitted to nor are they under consideration for publication by another journal, conference proceedings, or similar publication.

Information for Authors

Memoirs are printed from camera copy fully prepared by the author. This means that the finished book will look exactly like the copy submitted.

The paper must contain a *descriptive title* and an *abstract* that summarizes the article in language suitable for workers in the general field (algebra, analysis, etc.). The *descriptive title* should be short, but informative; useless or vague phrases such as "some remarks about" or "concerning" should be avoided. The *abstract* should be at least one complete sentence, and at most 300 words. Included with the footnotes to the paper should be the 2000 *Mathematics Subject Classification* representing the primary and secondary subjects of the article. The classifications are accessible from www.ams.org/msc/. The list of classifications is also available in print starting with the 1999 annual index of *Mathematical Reviews*. The Mathematics Subject Classification footnote may be followed by a list of *key words and phrases* describing the subject matter of the article and taken from it. Journal abbreviations used in bibliographies are listed in the latest *Mathematical Reviews* annual index. The series abbreviations are also accessible from www.ams.org/publications/. To help in preparing and verifying references, the AMS offers MR Lookup, a Reference Tool for Linking, at www.ams.org/mrlookup/. When the manuscript is submitted, authors should supply the editor with electronic addresses if available. These will be printed after the postal address at the end of the article.

Electronically prepared manuscripts. The AMS encourages electronically prepared manuscripts, with a strong preference for $\mathcal{A}_{\mathcal{M}}\mathcal{S}$-LaTeX. To this end, the Society has prepared $\mathcal{A}_{\mathcal{M}}\mathcal{S}$-LaTeX author packages for each AMS publication. Author packages include instructions for preparing electronic manuscripts, the *AMS Author Handbook*, samples, and a style file that generates the particular design specifications of that publication series. Though $\mathcal{A}_{\mathcal{M}}\mathcal{S}$-LaTeX is the highly preferred format of TeX, author packages are also available in $\mathcal{A}_{\mathcal{M}}\mathcal{S}$-TeX.

Authors may retrieve an author package from e-MATH starting from
www.ams.org/tex/ or via FTP to ftp.ams.org (login as **anonymous**, enter
username as password, and type cd pub/author-info). The *AMS Author Handbook* and the *Instruction Manual* are available in PDF format following the author
packages link from www.ams.org/tex/. The author package can be obtained free
of charge by sending email to pub@ams.org (Internet) or from the Publication Division, American Mathematical Society, P.O. Box 6248, Providence, RI 02940-6248.
When requesting an author package, please specify $\mathcal{AMS}$-LaTeX or $\mathcal{AMS}$-TeX, Macintosh or IBM (3.5) format, and the publication in which your paper will appear.
Please be sure to include your complete mailing address.

Sending electronic files. After acceptance, the source file(s) should be sent to
the Providence office (this includes any TeX source file, any graphics files, and the
DVI or PostScript file).

Before sending the source file, be sure you have proofread your paper carefully.
The files you send must be the EXACT files used to generate the proof copy that was
accepted for publication. For all publications, authors are required to send a printed
copy of their paper, which exactly matches the copy approved for publication, along
with any graphics that will appear in the paper.

TeX files may be submitted by email, FTP, or on diskette. The DVI file(s) and
PostScript files should be submitted only by FTP or on diskette unless they are
encoded properly to submit through email. (DVI files are binary and PostScript
files tend to be very large.)

Electronically prepared manuscripts can be sent via email to
pub-submit@ams.org (Internet). The subject line of the message should include
the publication code to identify it as a Memoir. TeX source files, DVI files, and
PostScript files can be transferred over the Internet by FTP to the Internet node
e-math.ams.org (130.44.1.100).

Electronic graphics. Comprehensive instructions on preparing graphics are available at www.ams.org/jourhtml/graphics.html. A few of the major requirements are given here.

Submit files for graphics as EPS (Encapsulated PostScript) files. This includes
graphics originated via a graphics application as well as scanned photographs or
other computer-generated images. If this is not possible, TIFF files are acceptable
as long as they can be opened in Adobe Photoshop or Illustrator. No matter what
method was used to produce the graphic, it is necessary to provide a paper copy to
the AMS.

Authors using graphics packages for the creation of electronic art should also
avoid the use of any lines thinner than 0.5 points in width. Many graphics packages
allow the user to specify a "hairline" for a very thin line. Hairlines often look
acceptable when proofed on a typical laser printer. However, when produced on a
high-resolution laser imagesetter, hairlines become nearly invisible and will be lost
entirely in the final printing process.

Screens should be set to values between 15% and 85%. Screens which fall outside
of this range are too light or too dark to print correctly. Variations of screens within
a graphic should be no less than 10%.

Inquiries. Any inquiries concerning a paper that has been accepted for publication should be sent directly to the Electronic Prepress Department, American
Mathematical Society, P. O. Box 6248, Providence, RI 02940-6248.

Editors

This journal is designed particularly for long research papers, normally at least 80 pages in length, and groups of cognate papers in pure and applied mathematics. Papers intended for publication in the *Memoirs* should be addressed to one of the following editors. In principle the Memoirs welcomes electronic submissions, and some of the editors, those whose names appear below with an asterisk (*), have indicated that they prefer them. However, editors reserve the right to request hard copies after papers have been submitted electronically. Authors are advised to make preliminary email inquiries to editors about whether they are likely to be able to handle submissions in a particular electronic form.

Algebra to CHARLES CURTIS, Department of Mathematics, University of Oregon, Eugene, OR 97403-1222 email: cwc@darkwing.uoregon.edu

Algebraic geometry and commutative algebra to LAWRENCE EIN, Department of Mathematics, University of Illinois, 851 S. Morgan (M/C 249), Chicago, IL 60607-7045; email: ein@uic.edu

Algebraic topology and cohomology of groups to STEWART PRIDDY, Department of Mathematics, Northwestern University, 2033 Sheridan Road, Evanston, IL 60208-2730; email: priddy@math.nwu.edu

Combinatorics and Lie theory to SERGEY FOMIN, Department of Mathematics, University of Michigan, Ann Arbor, Michigan 48109-1109; email: fomin@math.lsa.umich.edu

Complex analysis and complex geometry to DUONG H. PHONG, Department of Mathematics, Columbia University, 2990 Broadway, New York, NY 10027-0029; email: dp@math.columbia.edu

*__Differential geometry and global analysis__ to LISA C. JEFFREY, Department of Mathematics, University of Toronto, 100 St. George St., Toronto, ON Canada M5S 3G3; email: jeffrey@math.toronto.edu

*__Dynamical systems and ergodic theory__ to ROBERT F. WILLIAMS, Department of Mathematics, University of Texas, Austin, Texas 78712-1082; email: bob@math.utexas.edu

Geometric topology, knot theory, hyperbolic geometry, and general topoogy to JOHN LUECKE, Department of Mathematics, University of Texas, Austin, TX 78712-1082; email: luecke@math.utexas.edu

Harmonic analysis, representation theory, and Lie theory to ROBERT J. STANTON, Department of Mathematics, The Ohio State University, 231 West 18th Avenue, Columbus, OH 43210-1174; email: stanton@math.ohio-state.edu

*__Logic__ to THEODORE SLAMAN, Department of Mathematics, University of California, Berkeley, CA 94720-3840; email: slaman@math.berkeley.edu

Number theory to MICHAEL J. LARSEN, Department of Mathematics, Indiana University, Bloomington, IN 47405; email: larsen@math.indiana.edu

Operator algebras and functional analysis to BRUCE E. BLACKADAR, Department of Mathematics, University of Nevada, Reno, NV 89557; email: bruceb@math.unr.edu

*__Ordinary differential equations, partial differential equations, and applied mathematics__ to PETER W. BATES, Department of Mathematics, Brigham Young University, 292 TMCB, Provo, UT 84602-1001; email: peter@math.byu.edu

*__Partial differential equations and applied mathematics__ to BARBARA LEE KEYFITZ, Department of Mathematics, University of Houston, 4800 Calhoun Road, Houston, TX 77204-3476; email: keyfitz@uh.edu

*__Probability and statistics__ to KRZYSZTOF BURDZY, Department of Mathematics, University of Washington, Box 354350, Seattle, Washington 98195-4350; email: burdzy@math.washington.edu

*__Real and harmonic analysis and geometric partial differential equations__ to WILLIAM BECKNER, Department of Mathematics, University of Texas, Austin, TX 78712-1082; email: beckner@math.utexas.edu

All other communications to the editors should be addressed to the Managing Editor, WILLIAM BECKNER, Department of Mathematics, University of Texas, Austin, TX 78712-1082; email: beckner@math.utexas.edu.

Selected Titles in This Series

(*Continued from the front of this publication*)

680 **Joachim Zacharias,** Continuous tensor products and Arveson's spectral C^*-algebras, 2000
679 **Y. A. Abramovich and A. K. Kitover,** Inverses of disjointness preserving operators, 2000
678 **Wilhelm Stannat,** The theory of generalized Dirichlet forms and its applications in analysis and stochastics, 1999
677 **Volodymyr V. Lyubashenko,** Squared Hopf algebras, 1999
676 **S. Strelitz,** Asymptotics for solutions of linear differential equations having turning points with applications, 1999
675 **Michael B. Marcus and Jay Rosen,** Renormalized self-intersection local times and Wick power chaos processes, 1999
674 **R. Lawther and D. M. Testerman,** A_1 subgroups of exceptional algebraic groups, 1999
673 **John Lott,** Diffeomorphisms and noncommutative analytic torsion, 1999
672 **Yael Karshon,** Periodic Hamiltonian flows on four dimensional manifolds, 1999
671 **Andrzej Rosłanowski and Saharon Shelah,** Norms on possibilities I: Forcing with trees and creatures, 1999
670 **Steve Jackson,** A computation of δ_5^1, 1999
669 **Seán Keel and James McKernan,** Rational curves on quasi-projective surfaces, 1999
668 **E. N. Dancer and P. Poláčik,** Realization of vector fields and dynamics of spatially homogeneous parabolic equations, 1999
667 **Ethan Akin,** Simplicial dynamical systems, 1999
666 **Mark Hovey and Neil P. Strickland,** Morava K-theories and localisation, 1999
665 **George Lawrence Ashline,** The defect relation of meromorphic maps on parabolic manifolds, 1999
664 **Xia Chen,** Limit theorems for functionals of ergodic Markov chains with general state space, 1999
663 **Ola Bratteli and Palle E. T. Jorgensen,** Iterated function systems and permutation representation of the Cuntz algebra, 1999
662 **B. H. Bowditch,** Treelike structures arising from continua and convergence groups, 1999
661 **J. P. C. Greenlees,** Rational S^1-equivariant stable homotopy theory, 1999
660 **Dale E. Alspach,** Tensor products and independent sums of $\mathcal{L}_p$-spaces, $1 < p < \infty$, 1999
659 **R. D. Nussbaum and S. M. Verduyn Lunel,** Generalizations of the Perron-Frobenius theorem for nonlinear maps, 1999
658 **Hasna Riahi,** Study of the critical points at infinity arising from the failure of the Palais-Smale condition for n-body type problems, 1999
657 **Richard F. Bass and Krzysztof Burdzy,** Cutting Brownian paths, 1999
656 **W. G. Bade, H. G. Dales, and Z. A. Lykova,** Algebraic and strong splittings of extensions of Banach algebras, 1999
655 **Yuval Z. Flicker,** Matching of orbital integrals on $GL(4)$ and $GSp(2)$, 1999
654 **Wancheng Sheng and Tong Zhang,** The Riemann problem for the transportation equations in gas dynamics, 1999
653 **L. C. Evans and W. Gangbo,** Differential equations methods for the Monge-Kantorovich mass transfer problem, 1999
652 **Arne Meurman and Mirko Primc,** Annihilating fields of standard modules of $\mathfrak{sl}(2,\mathbb{C})^\sim$ and combinatorial identities, 1999

For a complete list of titles in this series, visit the
AMS Bookstore at **www.ams.org/bookstore/**.